Extracellular Microbial Polysaccharides

Extracellular Microbial Polysaccharides

Paul A. Sandford, EDITOR
U. S. Department of Agriculture

Allen Laskin, EDITOR
Exxon Research and Engineering Co.

A symposium co-sponsored by
the Division of Carbohydrate
Chemistry and the Division of
Microbial and Biochemical
Technology at the 172nd
Meeting of the American
Chemical Society,
San Francisco, Calif.,
August 30–31, 1976

ACS SYMPOSIUM SERIES **45**

AMERICAN CHEMICAL SOCIETY
WASHINGTON, D. C.   1977

SEP

CHEM

Library of Congress CIP Data

Extracellular microbial polysaccharides.
  (ACS symposium series; 45 ISSN 0097-6156)

  Includes bibliographical references and index.

  1. Microbial polysaccharides—Congresses.
  I. Sandford, Paul A., 1939-   . II. Laskin, Allen I.,
1928-   . III. American Chemical Society. Division of
Carbohydrate Chemistry. IV. American Chemical Society.
Division of Microbial and Biochemical Technology. V.
Series: American Chemical Society. ACS symposium
series; 45.

QR92.P6E97   660'.62   77-6368
ISBN 0-8412-0372-5   ACSMC 8   45 1–326

S·D
7/6/77
LM

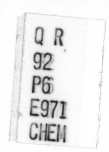

ACS Symposium Series

Robert F. Gould, *Editor*

FOREWORD

The ACS SYMPOSIUM SERIES was founded in 1974 to provide a medium for publishing symposia quickly in book form. The format of the SERIES parallels that of the continuing ADVANCES IN CHEMISTRY SERIES except that in order to save time the papers are not typeset but are reproduced as they are submitted by the authors in camera-ready form. As a further means of saving time, the papers are not edited or reviewed except by the symposium chairman, who becomes editor of the book. Papers published in the ACS SYMPOSIUM SERIES are original contributions not published elsewhere in whole or major part and include reports of research as well as reviews since symposia may embrace both types of presentation.

CONTENTS

PREFACE

A new fermentation industry, the production of extracellular microbial water-soluble polysaccharides, arose in the late 1950's and early 1960's and is now expanding rapidly. Several factors have accelerated the use of microbial polysaccharides as well as the search for new sources of water-soluble polysaccharides. Although hydrocolloids obtained from plants and seaweed have been used successfully for numerous applications in the food, textile, agricultural, paint, and petroleum industries, increasing labor costs, limited sources, adverse climate conditions, and increased demands have resulted in a constant or dwindling supply of several of these traditionally used plant and seaweed gums. Also industry has demands for water-soluble polymers that are not met by the traditional plant and seaweed gums.

Extracellular polysaccharide production is a widespread characteristic of microorganisms. Several of these polymers have proven to be commercially significant. The usefulness of these microbial polysaccharides primarily results from their unique physical and chemical properties which are determined by their individual component sugars and their mode of linkages. Their constant chemical properties and constant supply also increase their desirability. Other reasons for industry's interest in microbial gums are their potentially diverse sources and types.

This symposium focuses on the production and properties of extracellular microbial polysaccharides that are currently being used by industry or which have potentially useful industrial properties. Special emphasis is placed on new areas of research that would improve or stimulate industrial production and use of this valuable class of water soluble hydrocolloids.

U.S. Department of Agriculture   PAUL A. SANDFORD
Peoria, Ill. 61604

Exxon Research and Engineering Co.   ALLEN I. LASKIN
Linden, N.J. 07036
January 12, 1977

Culture Maintenance and Productivity

DENIS K. KIDBY

Department of Soil Science and Plant Nutrition, The University of Western Australia,
Nedlands, Western Australia, 6009

Microbial productivity is based upon a very large store of genetic information. In a typical bacterial cell, there are more than one million items encoded. At the initiation of inoculum build-up, it is a common practice to transfer approximately 10^9 cells to a fresh medium. To retain the complete genetic identity of such an inoculum, for even a single generation, 10^{16} base pairings must occur with complete fidelity. However, examination of such a cell population would reveal that thousands of errors had occurred. The fidelity of DNA replication is nevertheless impressive, and given skilful management, microbes can approach the reliability of solution chemistry in terms of product reproducibility. While genetic change may be a disaster when uncontrolled, it is also the means of improving productivity.

Genetic alterations were once achieved more or less by chance. However, the possibility now exists for the deliberate, and specific, alteration of genotype to yield productive chimeras limited only by the imagination. One can envisage the real possibility of producing a bacterial cell which could extract its energy and growth requirements from a few simple salts, the air and sunlight, producing a bacterial product such as Xanthan Gum or, an algal product such as agarose. However, despite such advances in the manipulation of genes, it seems certain that the inherent genetic instability of microbes will remain an important problem for many years; and it is largely to this type of difficulty that the present paper is addressed. Before discussing instability, the origins of industrial cultures will be briefly considered.

Sources of Microbes

Natural Sources. Many useful microbes are directly obtainable from the soil or other natural sources. It is often possible to employ unusual or extreme conditions as selective agents in the search for microbes with special abilities. Bacteria isolated from hot springs, can be grown near the temperature of boiling water (1). Acid mine leachings harbour

bacteria able to grow at high concentrations of sulfuric acid (2).
Microbes free of toxins or especially allergenic substances may
be sought in foodstuffs in which they are known to regularly
occur in high concentrations.

Isolation Procedures. The principles employed are those of
selective enrichment or inhibition. The required, or suspected,
nutritional and physiological characteristics of the organism
sought will dictate and actual procedure. The oxidation,
reduction, binding, or release, of dyes are particularly adap-
table for service as indicators of specific biochemical events.
The possession of a particular enzyme, or series of enzymes, may
be linked to either the ability, or inability, to grow on a
particular medium. Biological indicators such as the growth of
an indicator organism are particularly sensitive to such functions as the excretion of vitamins or amino acids. Ingenious
methods have been devised for the selection of characteristics
which are by their nature cryptic and seemingly inaccessible for
selection. For example Okanishi and Gregory (3) were able to
devise a simple method to reveal yeast colonies possessing
higher than normal methionine levels.

Protocols for the isolation of specific nutritional types
may be sought in the taxonomic literature (4, 5). Specific
procedures for various groups of organisms are available in the
recent literature (6, 7, 8). However, the seeker of desirable
microbes must often rely upon his own resourcefulness. A fairly
thorough biochemical understanding of the event of interest can
b e a most useful guide to isolation procedures.

In the case of extracellular products, such as polysaccharides, there may or may not be characteristically mucoid
colonies. Selective procedures should, if possible, exploit
some specific property of the desired polysaccharide. However,
there are possibilities for indirect selection using associated
characteristics. For example, many characteristics, suited to
replica-plating methods, are associated with polysaccharide
producing Xanthomonas campestris (9). In the case of mucoid
Escherichia coli, there appears to be associated UV sensitivity
(10). Replica-plating procedures are frequently the most useful
technique since one can select for cells which either grow or do
not grow. Diagnostic procedures which are destructive may also
be used since all material under investigation is retained on
the replicas. The employment of specific enzymes for the recognition of certain types of polysaccharides is an interesting
possibility for the development of screening programmes. In this
connection it is interesting to note that recognition systems
based upon enzyme specificity may already occur in bacteriophage
(11).

Culture Collections. Searching for microbes in existing
cultures will frequently be quicker, cheaper and easier than
isolation from nature. As an aid to such a search, Hesseltine
and Haynes (12) have written a guide to collections containing

industrially useful microbes. However, there can be no substitute
for thorough searching of the current literature.

Maintenance of Genotype

Nature of the Problem. An industrially useful microbe is an
asset which may range from being moderately valuable to almost
priceless. The preservation of such an asset deserves a priority
which it seldom receives. The greatest barrier to successful
preservation of genotype may be a failure to appreciate that:
(i) microbes are inherently unstable, (ii) there is no method yet
devised for the complete preservation of genotype.

Inherent Instability of Microbes. The potential for genotype variability has been indicated in the introductory remarks.
It is now necessary to discuss the actual mechanism of change and
how these relate to phenotype.

All regions of a gene are mutable. Some genes are more
mutable than others because they have intragenic regions of high
mutability, are influenced by some other gene which is itself
mutable or, are under the control of genes which promote mutation.
All of these mechanisms are known to occur, including some in
which the mutability is effected by an extrachromosomal element
or, an infectious agent (13). It is these more highly mutable
genes, and especially those cases involving infectious agents,
that are most troublesome. Mutations may be either replication-
dependent or replication-independent. It is speculated (14) that
replication-dependent mutations reflect errors in DNA replication,
and replication-independent mutations reflect error-prone repair
systems. Mutations may involve: (i) frame-shift; (ii) deletion;
(iii) insertion; (iv) base pair substitution. The effect on the
code may be either the production of missense, nonsense, or a
non-code function may be lost. The resulting phenotypes may
include: (i) altered RNA base sequence; (ii) altered amino acid
sequence; (iii) premature termination; (iv) degenerate silence.
A certain proportion of these mutants will be cryptic, particularly those involving missense. Mutations which lead to the
insertion of a similar amino acid or, because of code degeneracy,
the wild type amino acid, will usually not be revealed. It has
been calculated that 25% of 549 base pair substitutions involve
degeneracy (15). It is also interesting to note that there is a
greater than random probability that base pair substitutions will
lead to substitution of a similar rather than a dissimilar amino
acid (16). Lethal mutations will also be cryptic since these
will not persist, unless they are conditional.

Intragenic mutations are non-random. Sites which are highly
mutable are hot spots (17). Evidence on the nature of hot spots
has been reviewed by Clarke and Johnston (1976) and will be
merely summarized here.

High Mutability Regions. (i) Frameshift mutations tend to
occur in regions of repeated base pairs. Runs of either AT or
GC base pairs have been associated with frameshifts. (ii) Base

pair substitutions are influenced by neighbouring bases. The
AT-GC substitution induced by 2-aminopurine at the second position
of a triplet has been demonstrated to occur 23 times more frequently when an AT base pair was present in the third position
(18). (iii) Mutator polymerase acts preferentially on specific
regions of the gene. (iv) The frequency and location of deletions
is non-random and such sites are considered deletion hot spots.
(v) Ultra-violet induced mutations are most frequent in tracts
of pyrimidines.

Development of a Stable Mutation. Most mutations are formed
from pre-mutational lesions. The lesion may or may not be repaired or, the repair process itself may lead directly to mutation. Failing repair, the pre-mutational lesion may be established as a mutation by DNA replication. The events involved in
development of a mutation are summarized in Figure 1. Any one of
these steps may be subject to the influence of adjacent base
pairs.

In the light of these observations, one might ask what
avenues exist for the amelioration or removal of hot spots? If
the mutation is effected by a mutagen, it may be possible to
either remove or suppress the condition leading to the presence
of the mutagen or neutralize its activity with an antimutator.
Precedents for this latter approach are now well documented (14).

Antimutagenesis. It has been quite properly stated (14)
that one cannot understand mutagenesis or the regulation of
mutation frequency without considering antimutagenic effects.
Antimutagenesis may be defined as a decrease in the actual rate
of mutation. Decreased apparent rates may be caused by either
altered survival or dose reduction, and these effects are termed
apparent antimutagenesis. A mutation or premutation may arise by:
(i) reaction between a mutagen and DNA; (ii) incorporation of a
mutagen-altered precursor or base analogue; (iii) replication
error; (iv) recombination error; (v) repair error; (vi) transcription error; (vii) translation error. The last two
mechanisms involve the production of error-prone RNA or proteins
which alter the base sequence of DNA either directly or indirectly (19, 20, 21).

Clarke and Shankel (14) have distinguished between genetic
antimutagenesis, which is the antimutagenic effect of replication
genes, repair genes, or other genetic determinants, and physiological antimutagenesis which is achieved by added chemicals or
altered cell conditions. The physiological mechanism would
appear to offer considerable potential for the reduction of
mutation rates for certain classes of mutation. For example,
adenosine appears to be capable of virtually abolishing the
mutagenicity of purine mutagens (14). Spontaneous mutation rates
have also been dramatically reduced by the use of acridines (22).
An observation of considerable interest is that genes are more
likely to mutate when being transcribed (14). Thus the repression of gene activity is antimutagenic. It might be expected,

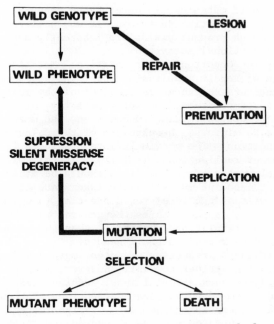

Figure 1.   *Sequences of events in mutation and selection*

therefore, that in maintenance and inoculum build-up cultures, the repression of the productive function would help to arrest variability by decreasing the rate of mutation.

It may also be the case that repression of product formation will help prevent selection against producer cells. There is some evidence (23, 24) that product repression may be of use in reducing variability in Xanthomonas campestris. There seems little reason to doubt that DNA which is not being transcribed should be relatively stable. It would be of considerable interest to see if mutations in derepressed genes are in fact proportional to transcription rates. It may well be that certain microbes with high product yields are inherently unstable because of high transcriptional activity.

Limiting the Opportunity for Mutation. Mutation rates may be a function of repair, replication or translation rates, of mutagen or antimutagen concentrations, or of physical conditions such as raised temperature, low water activity, or ice crystals. Whatever the condition leading to mutation, the most effective protection is to minimise the exposure of the culture to the conducive condition. The growth in mutant numbers is a function of the number of replications (Table I). It follows, therefore, that the total number of replications should be minimized. If replication-independent mutations are taken into account, then it also follows that the total residence time in culture should be minimized. If, as seems to be the general case, mutation is proportional to translational activity, then the productive function should be repressed until needed.

The exclusion or reduction of potent mutagens may seem too obvious to require further comment. However, many commonly occurring mutagens such as metal ions, adenine, caffeine, ozone, to name a few, seem often to escape attention. The number of base analogues generated by chemical, or high temperature, treatment of concentrated sources of purine and pyrimidine bases must often be considerable. The frequent proximity of cultures to electric motors and, in particular, atmospheres recently irradiated with ultra-violet light must surely produce large numbers of ozone-induced mutants. Extremely high levels of mutation have been observed in E. coli exposed to as little as 0.1 ppm ozone for 60 minutes (10).

The question of limiting the opportunity for mutation will be further discussed in connection with preservation techniques.

Limiting the Opportunity for Selection. The selection of a mutant, in the present context, may be taken to mean the increase of any given mutant to a significant proportion of the total population. The extent of this selection will be a function of the culture conditions and the number of generations of culture growth permitted. Selective media may be employed to remove particular classes of mutant. Nutritionally rich media will tend to preserve and often concentrate auxotrophs while a poorer medium may select fairly efficiently against auxotrophs, unless

TABLE I

THE PROPORTION OF MUTANTS IN A GROWING CULTURE

Generations	0	1	2	3	4
Total Cells	N	2N	4N	8N	16N
Mutant Cells[a]	0	2mN	8mN	24mN	64mN
Mutant: Total	0	m	2m	3m	4m

[a] m = mutation rate

high rates of cross-feeding occur.

Short-term Preservation. The preservation of cell viability
for periods of less than a few months might arbitrarily be termed
short-term preservation. While there can be no doubt as to the
desirability of long-term preservation, methods of achieving this
usually provide relatively inaccessible inocula and, in some
cases, may be of limited success. In order to be useful, a short
-term preservation method must provide a high recovery of viable
cells which grow with a minimum lag phase. The inoculum should
be easily accessible and of a standard and suitable size. Sub-
culture to achieve vigorously growing and reproducible cultures
should not be necessary. If these criteria cannot be met, it may
be better to consider the routine use of inocula preserved by
long-term methods.

Useful short-term preservation methods are generally variations of drying procedures. A particularly suitable method is
the drying of cultures onto paper (25, 26). Paper strips have
the advantage of being easily handled and are readily adjusted in
size to yield an appropriate inoculum size. The method has been
used with success for X. campestris NRRL B1459 (9). Other short-
term preservation methods have been reviewed elsewhere (26).

The repeated transfer of cultures for routine maintenance
must be considered an unwise practice and is difficult to justify
where alternative non-propagative methods exist.

Long-term Preservation. Storage of lyophilized, frozen, or
L-dried cells are the principle means of long-term preservation
(26). There is an extremely widespread belief that the method of
choice is lyophilization. This belief is not justified by either
fact or theory.

The reasons for the widespread preference for lyophilization
are: (i) this was the first generally successful method of long-
term preservation; (ii) the product has an "attractive" appearance; (iii) injury from concentrated solutes in the liquid state
seemed a reasonable supposition; (iv) protection against injury
by drying at freezing temperatures seemed an attractive advantage.
It is now clear that highly concentrated solutes are not as injurious as has been formerly supposed and may in fact exert significant protection (27). In the light of extensive investigations of the L-drying methods of Annear (28-33) by other workers
(26, 34, 35), it seems that this procedure is to be preferred
since recovery of many difficult to preserve organisms is
typically 10 to 100 times higher than is achieved with lyophilization. It has also been observed that large increases in
mutants can accompany lyophilization (36, 37, 38). While no
proper comparison appears to have been made between mutant yields
from lyophilization and L-drying, it seems reasonable to expect
that the higher recoveries obtained by L-drying would be
accompanied by less damage and therefore fewer mutants.

There are a number of steps in preservation and subsequent
recovery procedures which may cause genetic damage (Figure 2).

Freezing is in itself injurious (39). The extent of drying also
appears to influence the yield of mutations (40, 41). Prophage
may also be induced by desiccation (42, 43). The rehydration
procedure is also of importance and there appears to be some
evidence of cell leakage leading to poor recovery (27). The recovery medium is an important selective agent and can clearly
influence the recovery of certain types of mutants. For example,
some medium components can inhibit recovery of nonsense suppress-
ors in Saccharomyces cerevisiae, while other components can relieve this inhibition (44).

Storage in the frozen state has little to recommend it except convenience. Storage itself is not considered to be injurious provided that ice crystal damage is precluded by holding
the temperature below -130ºC (45).

It is suggested that for preservation of genotype, L-drying
procedures for both long and short-term requirements may be found
particularly successful.

It is not clear how low a temperature should be employed for
storage of dried material, but in the absence of evidence to the
contrary, as low a temperature as is available would seem desirable. For long-term preservation, the material is normally
held under vacuum while for short-term preservation, less stringent, and therefore more convenient, conditions may be employed.
When rehydrating, a low cell:culture volume ratio should be
employed. The culture medium should be as nutritionally rich as
is consistent with good growth. This procedure will to some
degree select for auxotrophs. However, it is possible to screen
these out in subsequent culture if necessary. No cell population
is genetically identical to its parent culture. The change in
identity can, however, be minimized by the use of methods which
lead to high recovery rates. The preservation of fresh isolates
should not be delayed and it is worth adopting a standard protocol
to deal with this situation (Figure 3).

Improvement of Genotype.
    Control Mutants. One of the most useful types of mutant is
the control mutant where feed-back inhibition or repression is
absent. In the case of polysaccharide production such mutants
are most likely to be recognized by their production of large
mucoid colonies.
    Conditional mutants. The conditional mutant has great poten-
tial for controlling complex cell functions by such simple means
as raising or lowering of temperature. Such mutants are relatively easy to obtain. For example, polysaccharide production which
is conditional may be switched on and off or, conditional growth
may be switched off to permit polysaccharide production in the
absence of growth. Conditional lysis is also of considerable
application where it is desirable, and it usually is, to remove
the cells from the completed fermentation. Lysis may be
achieved by the induction of bacteriophage. Bacteriocins also

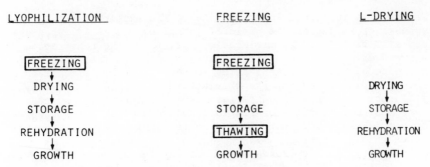

Figure 2.   *Comparisons between sequences of events involved in preservation of cells and their subsequent recovery*

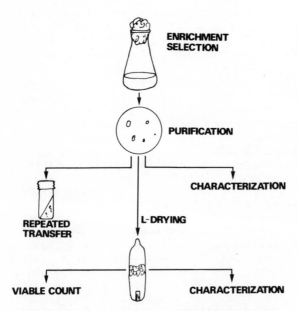

Figure 3.   *Selection and preservation of microbes. The scheme described incorporates tests of L-dried cultures to determine viability and any alteration of characteristics as a result of the preservation procedure.*

offer great potential for lysing of cultures.

Stabilized Genes. The potential for stabilization varies
according to the origin of the instability. In the case of hot
spots, the breaking up of runs of base pairs might be expected to
be effective. An increase in the number of genes may be effective
and may, if translation is the rate-limiting step in production,
also lead to higher production levels. It may be possible to
transfer genes from a related organism exhibiting a more stable
genotype. Stable genotypes may be fairly readily revealed by
employing the selective pressure of chemostat culture (46).

Methods for Genotype Alteration. Genotypes are altered by:
(i) induced mutation; (ii) spontaneous mutation; (iii) transfer
of existing genes. The first method is rapid and some degree of
specificity is possible as for example in the case of ozone and
UV induced mutants (10). However, a large background of unwanted mutations may also be present. Spontaneous mutation rates
are, of course, slower, but are capable of producing the required
mutants in a surprisingly short time. The selection pressure to
obtain particular types of spontaneous mutants should be applied
in a continuous, rather than a discontinuous, manner. This
permits a more complete range of possibilities to be expressed
and is likely to lead to a more stable mutant since the desired
character can be acquired by a series of small steps rather than
one large step which could, for example, be due to a single point
mutation. For example, stable and high level antibiotic resistance has been achieved in Xanthomonas by using gradient plates
but was not readily achieved when using discrete steps (24). A
particularly helpful account of methods of mutant isolation is
given by Hopwood (47).

Perhaps the most attractive methods of genotype improvement
involve transfer of genetic material. The advantage of this
method is specificity, stability, and relative freedom from unwanted changes in other genes. Some very exciting alterations
can be attempted by this means. It is desirable for the organisms to be closely related because the transferred gene is more
likely to behave characteristically in the recipient. However,
genes certainly are transferable between distantly related
species and genetic engineering may be expected to revolutionize
the synthesis of natural products.

The methods of genetic transfer among bacteria are: (i)
conjugation; (ii) transduction; (iii) transfection; (iv) transformation, and (v) in vitro recombination and transfer from
divergent species or genetic engineering. The first four
methods are conventional and are extensively described (48).
However, genetic engineering is a combination of methodologies
and the total procedure may be varied considerably. One recently
described method (49) consists of isolation of the gene as its
RNA transcription product, retranscription back to DNA and synthesis of a complementary strand. These strands are elongated
with homopolymer tails of oligo-(dG). This double stranded gene

is then mixed with a plasmid which has been prepared as follows.
A nick is placed in the circular plasmid to provide linear DNA
which is repaired then extended with a homopolyer tail of oligo-
(dC) which is, of course, complementary to the artificial tail on
the copied gene. The plasmid picks up the gene by the complementary tail sections and, in doing so, becomes circular and thus
infective. Following infection, the plasmid is covalently linked
to the copied gene by host enzymes. This gene may be transferable
to a wide range of bacteria.

Furthermore, in this particular example, the gene may be
removed again from the plasmid, using a specific restriction
nuclease, and transferred to some other plasmid.

Thus it is possible to conceive of natural products which
are either inaccessible or grown on a seasonal basis, growing in
fermenters within hours. This has considerable implications, not
only for production costs but for the relative ease with which
production volumes can be regulated.

Abstract
      Sources of microbes and procedures for their selection,
isolation and maintenance are discussed. Maintenance of genotype
is considered in terms of the nature of genetic variability,
antimutagenesis, inoculation schedules, growth media and preservation methods.

      The improvement of genotype is discussed in terms of control
mutants, conditional mutants, and methods of genotype alteration.
Some common practices which may be conducive to culture
degeneration are discussed and suggestions are made as to
alternative procedures.

Literature Cited
  1. Brock, T.D., Ann. Rev. Ecology System (1970) 1, 191.
  2. Lundgren, D., et al., "Water Pollution Microbiology", John
      Wiley, New York (1972) 69-88.
  3. Okanishi, M., Gregory, K.F., Canad. J. Microbiol. (1970) 16,
      1139.
  4. "Bergey's Manual of Determinative Bacteriology" Williams and
      Wilkins.
  5. "Abstracts of Microbiological Methods", John Wiley, New
      York (1969).
  6. "Methods in Microbiology" 3A, Academic Press, New York (1970)
  7. "Methods in Microbiology" 3B, Academic Press, New York (1970)
  8. "Methods in Microbiology" 4, Academic Press, New York (1971)
  9. Kidby, D.K., et al., unpublished.
 10. Hamelin, C., Chung, Y.S., Mutat. Res. (1975) 28, 131.
 11. Sutherland, I.W., J. gen. Microbiol. (1976) 94, 211.
 12. Hesseltine, C.W., Haynes, W.C., Progress in Industrial
      Microbiology (1973) 12, 3.
 13. Clarke, C.H., Johnston, A.W.B., Mutat. Res. (1976) 36, 147.
 14. Clarke, C.H., Shankel, C.M., Bacteriol. Rev. (1975) 39, 33.

15. Drake, J.W., "The Molecular Basis of Mutation", Holden-Day, San Francisco, 1970.
16. Vogel, F., J. Molec. Evoln. (1972) 1, 334.
17. Benzer, S., Proc. Natl. Acad. Sci. (1961) 47, 403.
18. Koch, R.E., Proc. Natl. Acad. Sci. (1971) 68, 773.
19. Lewis, C.M., Tarrant, G.M., Mutat. Res. (1971) 12, 349.
20. McBride, A.C., Gowans, C.S., Genet. Res. (1969) 14, 121.
21. Talmud, P., Lewis, D., Nature (1974) 249, 563.
22. Puglisi, P.P., Mutat. Res. (1967) 4, 289.
23. Cadmus, M.C., et al., Can. J. Microbiol. (1976) in press.
24. Kidby, D.K., unpublished.
25. Coe, A.W., Clark, S.P., Mon. Bull. Minist. Hlth. (1966) 25, 97.
26. Lapage, S.P. et al., "Methods in Microbiology" 3A, Academic Press, New York, (1970) 167.
27. Leach, R.H., Scott, W.J., J. gen. Microbiol. (1959) 21, 295.
28. Annear, D.I., Nature (1954) 174, 359.
29. Annear, D.I., J. Hyg. Camb. (1956) 54, 487.
30. Annear, D.I., J. Path. Bact. (1956) 72, 322.
31. Annear, D.I., J. Appl. Bact. (1957) 20, 17.
32. Annear, D.I., Aust. J. exp. Biol. med. Sci. (1958) 36, 1. .
33. Annear, D.I., Aust. J. exp. Biol. med. Sci. (1962) 40, 1.
34. Hopwood, D.A., Ferguson, H.M., J. appl. Bact. (1969) 32, 434.
35. Muggleton, P.W., Progr. Ind. Microbiol. (1962) 4, 191.
36. Hieda, K., Ito, T., "Freeze-drying of biological Materials" International Institute of Refrigeration, Paris (1973) 71.
37. Webb, S.J., Tai, C.C., Canad. J. Microbiol. (1968) 14, 727.
38. "Cryobiology", Academic Press, N.Y. (1966) 213.
39. Mazur, P., Science (1970) 168, 939.
40. Webb, S.J., Nature (1967) 213, 1137.
41. Webb, S.J. and Dumasia, M.D., Canad. J. Microbiol. (1968) 14, 841.
42. Webb, S.J. and Dumasia, M.D., Canad. J. Microbiol. (1967) 13, 33.
43. Webb, S.J. and Dumasia, M.D., Canad. J. Microbiol. (1967) 13, 303.
44. Queiroz, C., Biochem. Genet. (1973) 8, 85.
45. Martin, S.M., Ann. Rev. Microbiol. (1964) 18, 1.
46. Veldkamp, H., "Methods in Microbiology" 3A Academic Press, New York (1970) 305.
47. Hopwood, D.A., "Methods in Microbiology" 3A Academic Press, New York (1970) 363.
48. Hayes, W., "The Genetics of Bacteria and their Viruses" Blackwell, Oxford (1968).
49. Rougeon, F., Kourilsky, P., Mach, B., Nucleic Acids Res. (1975) 2, 2365.

2

The Production of Alginic Acid by *Azotobacter vinelandii* in Batch and Continuous Culture

L. DEAVIN, T. R. JARMAN, C. J. LAWSON, R. C. RIGHELATO, and S. SLOCOMBE

Tate & Lyle Ltd., Group Research and Development, Philip Lyle Memorial Research Laboratory, P.O. Box 68, Reading, Berks., RG6 2BX, U.K.

The production of polysaccharides by fermentation has been heralded by some of the more optimistic microbial technologists as the next major fermentation area. It is now gaining similar treatment in public and private meetings to that offered to single cell protein some years ago. This optimism is based on the undoubted success of the one major product, xanthan gum, which has raised the tantalising prospect of a whole range of microbial gums which would not only reflect and improve upon the available plant gums, but also introduce novel properties for exploitation in existing and as yet undeveloped applications. About a dozen companies are thought to be developing on a large scale the production of microbial polysaccharides; some of them are already in the fermentation industry but others, like our own, are newcomers to this technology. Despite this enormous research and development effort the state of the technology, as judged from patents and the scientific literature, is relatively poorly advanced. There is little public literature on the production technologies used by industry and academic microbiology has for the most part ignored the physiology of exocellular polysaccharide synthesis and excretion.

For this reason, we, along with other groups, have been studying the physiology of polysaccharide synthesis as a basis for developing production processes. In order to gain a greater understanding of the effects of individual environmental parameters on cell growth and polysaccharide synthesis continuous flow cultures(1) have been used wherever possible. For those unfamiliar with the methods of mass cultivation of microbes, the time honoured industrial and laboratory method is to inoculate a small amount of the microbe into a medium containing all of the necessary nutrients for growth and product formation. The microbes then grow until one or other substrate is exhausted and then growth stops. This is a simple batch culture system.

In continuous flow culture, by contrast, the nutrient medium is continuously added to the culture and the culture continuously harvested.

The ratio of the flow rate of the medium to the culture volume is called the dilution rate, and except at the maximum growth rate of the microbe, the concentration of one of the substances in the medium determines the concentration of the microbes. This is called the growth-limiting substrate. It is well established that changes in growth-limiting substrate can considerably affect the physiology of microbes. So too can changes in the dilution rate, which in a steady state is equal to the specific growth rate. In continuous cultures steady states can be maintained indefinitely and changes in individual parameters can readily be studied. By contrast in batch cultures, concentration of nutrients, cells and products, and all of these with respect to cell age, change continuously, which makes the study of cell physiology and biochemistry extremely complicated. This is illustrated by some batch fermentation processes for exopolysaccharides. The best known is of course xanthan production by Xanthomonas campestris.

In the simplest fermentation described by Moraine and Rogovin (2), the concentrations of the major substrates change throughout the fermentation. So too do the main products: bacterial cells and polysaccharide. Analysis of several batch cultures led Moraine & Rogovin (2) to conclude that several factors, including xanthan concentration, affected the rate of xanthan production, though the details of the relationship were not clear. The complicated kinetic pattern that emerged from these studies has been of considerable value in understanding the batch fermentation process for xanthan gum but does not enhance the understanding of the control of bio-synthesis, as it necessarily deals primarily with the effect of the changing fermentation parameters on the environment of the cells rather than directly with the effect of the environment on the cells.

In batch cultures of a Pseudomonas sp. which produces an exopoly-saccharide composed of glucose and galactose in the ratio 7 : 1 and contains both acetate and pyruvate (3) polymer synthesis was detectable in the later part of the exponential growth phase (Figure 1) and continued maximally during the period of zero specific growth rate, the so-called stationary phase (4). The limiting substrate, that is the substrate which determined the cell mass that was finally obtained, was not established in these cultures.

Another example of batch cultivation for an exopolysaccharide is that of alginic acid production by Azotobacter vinelandii. When the organism was grown under phosphate-deficient conditions polysaccharide synthesis continued throughout the growth phase but in contrast to the last example ceased when the microbes stopped growing (Figure 2).

From the studies of batch cultures of the types discussed it is difficult to draw any conclusions on the way in which bacteria control the synthesis of these exopolysaccharides. It has been supposed by many microbiologists that such products would be formed when a cell has an excess of carbohydrate

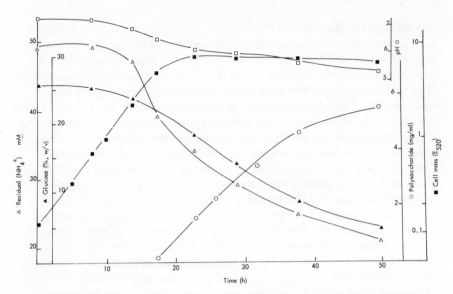

Figure 1.   Exopolysaccharide production by Pseudomonas sp. *in batch culture (From: Williams, A. G. (1974). Ph.D. Thesis, University College, Cardiff)*

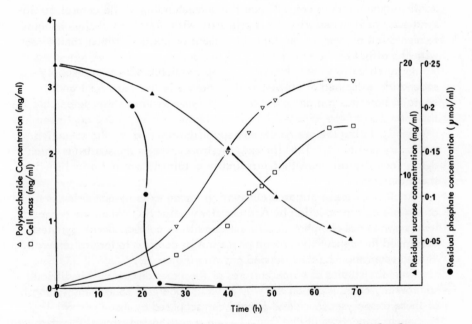

Figure 2.   Production of alginic acid by Azotobacter vinelandii *in batch cultures*

substrate and its growth is restricted by some other parameter. Neijssel and Tempest (5) have recently suggested from studies of Aerobacter aerogenes that they act as ATP sinks and are produced maximally under conditions which would cause the cells to overproduce ATP, conditions such as nitrogen limitation. The observations on Azotobacter vinelandii would perhaps contradict that particular hypothesis since high production rates were observed under phosphate-deficient conditions (Figure 2). Measurement of the rates of synthesis under a variety of environmental conditions might shed some light on the cellular control and the role of exopolysaccharide production. The major rate controlling process in a cell is its specific growth rate. A complex network of control mechanisms exist which permit the microbe to assimilate substrates, synthesise intermediates and form polymers (i.e. proteins, nucleic acids, cell walls, etc.) at rates which produce more cellular material of the same type and in similar ratios in the face of enormous environmental changes. It seems logical, then, to look first at the effect of growth rate on exopolysaccharide synthesis in continuous culture systems.

Silman and Rogovin (6) studied continuous cultures of Xanthomonas campestris in cultures thought to be limited by the nitrogenous component in the medium. pH was not controlled in these experiments so the data has been redrawn taking only the conditions in which the pH was between 6.3 and 7.2, a range in which it has been found that pH has little effect on xanthan production (Figure 3). At growth rates between 0.05 and 0.20 h^{-1} i.e. doubling times between 14 and 3.5 h, there was little change in the specific rate of synthesis of xanthan. The concentration of xanthan therefore increased with decreasing dilution rate. It is interesting to note that the xanthan production rate in these cultures varied only 15% either side of the mean value. This is quite different from the batch culture analysis which showed a threefold change in specific rate of xanthan production over a similar concentration range (2).

A similar independence of the rate of exopolymer synthesis on specific growth rate was found both with the Pseudomonas polysaccharide (4) and alginic acid synthesis by Azotobacter vinelandii. Over an even wider growth rate range the specific rate of synthesis of Pseudomonas exopolymer varied only 25% about the mean (Figure 4), whilst the polysaccharide concentration increased in proportion to the residence time of the culture (the residence time is the reciprocal of the dilution rate). In phosphate-limited continuous cultures of Azotobacter vinelandii the rate of alginate synthesis was independent of specific growth rate (Figure 5). In this case there was an increase in biomass at lower dilution rates. This was almost entirely due to the intracellular accumulation of the storage compound poly-B-hydrosybutyrate. With these three polysaccharides, then the rate of synthesis appears to be independent of the rate of growth and hence independent of the rate of most of the other intracellular biosyntheses.

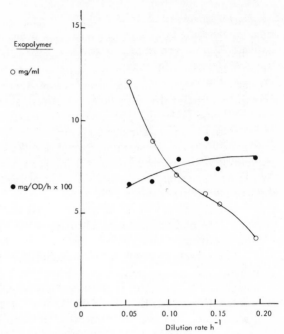

Biotechnology and Bioengineering

Figure 3. Effect of dilution rate on the production of xanthan by Xanthomonas campestris *in continuous culture (6)*

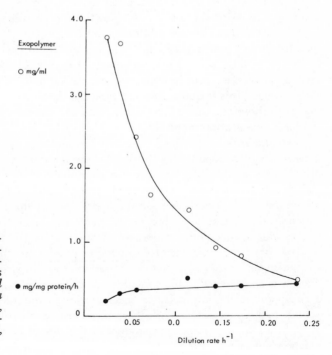

Figure 4. Effect of dilution rate on production of an exopolysaccharide by Pseudomonas sp *in ammonia-limited continuous culture (Data from Williams A. G., 1975; Ph.D. Thesis University College, Cardiff, U.K.)*

We have studied alginic acid synthesis by Azotobacter vinelandii in some detail and would like to pursue this argument with that particular system. Alginate as obtained from the conventional source, the brown algae, is a 1,4-linked linear copolymer of β-D-mannuronic acid and its 5-epimer α-L-guluronic acid (7) (Figure 6). The arrangement of monomers in this copolymer has been referred to as the block structure (8), the polymer having been shown to consist of regions of homo-polymeric blocks of mannuronic acid and of guluronic acid together with the so-called alternating or random sequences. The properties of the polymer, especially with respect to its gelling in the presence of calcium ions, depends both on the mannuronic acid to guluronic acid ratio and the block structure, the higher the proportion of polyguluronic acid blocks in the polymer the stronger and more brittle the gel formed in the presence of calcium ions (9). The polymer produced by Azotobacter vinelandii has the same basic structure as that from algal sources except that it is partially acetylated, approximately one in ten of the C2 and/or C3 hydroxyl groups being esterified with acetate (10, 11).

The markets for alginates demand products having a range of solution viscosities and gelling qualities. A range of alginate types comparable with algal products can be produced by Azotobacter vinelandii by appropriate choice of fermentation conditions. Haug and Larsen (12) showed that the mannuronic to guluronic acid ratio of Azotobacter alginate could be influenced by the calcium ion concentration of the growth medium and they presented evidence which suggested that mannuronic acid residues were epimerised to guluronic acid residues by an extracellular enzyme dependent on calcium ions for activity. In addition we have been able to manipulate the molecular weight and thus solution viscosity of the product produced by Azotobacter vinelandii.

By appropriate choice of fermentation conditions products with a wide range of viscosities were obtained which compared favourably with certain commercial algal alginates having low, medium and high viscosities (Figure 7). The results reported here apply to products obtained from continuous cultures but products with a similar range of viscosities can also be obtained from batch cultures.

The metabolism of Azotobacter vinelandii in relation to polysaccharide biosynthesis is shown in Figure 8. Sucrose, the carbohydrate growth substrate used, is transported into the cell, inverted, and glucose-6-phosphate and fructose-6-phosphate formed by their respective kinases. Fructose-6-phosphate enters the alginate biosynthetic pathway which has been shown to be via mannose-6-phosphate, mannose-1-phosphate and GDP-mannose nucleotide which is oxidised to GDP-mannuronic acid (13). Mannuronic acid residues are then polymerised to form polymannuronate which is partially epimerised extracellularly (12), to yield alginate. Azotobacter is an obligate aerobe, carbohydrate growth substrates are metabolised via

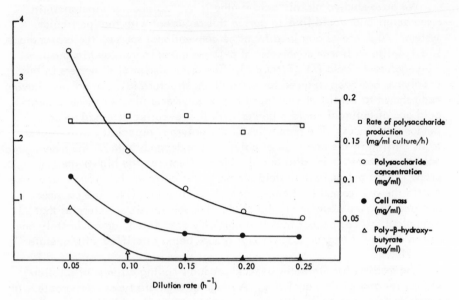

Figure 5. *Exopolysaccharide production by* Azotobacter vinelandii *at a range of dilution rates*

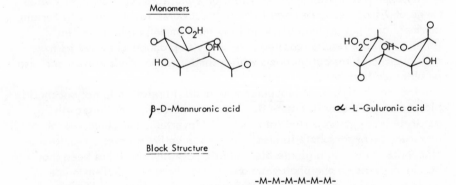

Figure 6. The structure of alginic acid

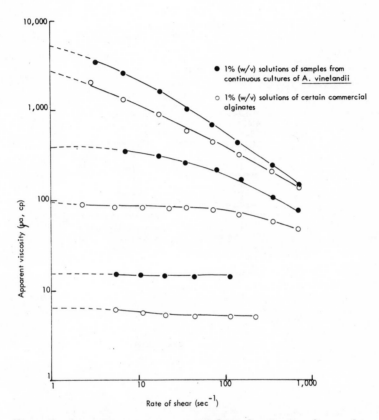

*Figure 7.   Apparent viscosity vs. rate of shear plots for Azotobacter algi-
nates and certain commercial algal alginates*

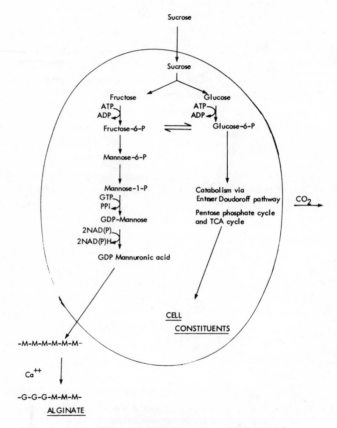

Figure 8.   Metabolism of Azotobacter vinelandii *in relation to alginate synthesis*

the Entner-Doudoroff pathway, pentose phosphate cycle and tricarboxylic acid cycle (14) and are oxidised to carbon dioxide. The products of sucrose metabolism are essentially alginate, biomass and carbon dioxide. With increases in oxygen tension Azotobacter exhibit an increase in respiration rate (15); the efficiency of energy conservation falling until at very high respiration rates as much as 90% of the sucrose utilised can be burnt off as carbon dioxide. One of the problems in developing a process for Azotobacter alginate production has therefore been to control this adverse respiration. This was a difficult proposition in batch culture with continually changing biomass and oxygen demand, especially as oxygen limitation has proved to be a disadvantageous condition for polymer production. Trials in batch culture under phosphate-deficient conditions indicated maximal obtainable yields of sodium alginate to be approximately 25% of the sucrose utilised. The effect of respiration rate on alginate production in continuous cutlure was therefore investigated.

The organism was grown at a range of specific respiration rates obtained by altering the fermenter impeller speed thus changing the rate of oxygen transfer into the culture broth (Figure 9). We chose phosphate-limited growth conditions, as a phosphate deficient medium, as discussed earlier was known to be conducive to polysaccharide synthesis in batch culture. Although cell mass, which remained essentially constant,was limited by availability of phosphate, the specific respiration rate was determined by oxygen availability. Polysaccharide concentration was also essentially constant, decreasing only at very low respiration rates. The rate of alginate synthesis was therefore largely independent of both the rate at which sucrose entered the cell, as indicated by the amount of sucrose utilised, and the rate at which intermediates entered the catabolic pathways and were respired to carbon dioxide. The maximum yield of sodium alginate, which occurred at the lower respiration rates, was in the region of 45% of the sucrose utilised as compared with the yields of 25% obtained in batch culture. At higher respiration rates the yield fell dramatically due to a greater proportion of the sucrose being oxidised to carbon dioxide.

The effect of different growth limitations on alginate production has also been investigated. Steady state continuous cultures were obtained with different nutrients limiting growth but cell mass and also specific respiration rate were controlled to within narrow ranges. Polysaccharide, determined as isopropanol precipitated material,was produced under all limitations tested (Table 1). Molybdate limitation followed by phosphate limitation, the condition routinely used, gave the most favourable specific rates of polysaccharide synthesis. Surprisingly, under sucrose limitation, a condition where the cell would be expected to make the most efficient use possible of its available carbon and energy substrate, polysaccharide was still produced at similar rates to other limitations. It is difficult to compare oxygen limitation, one condition tested where the specific rate of

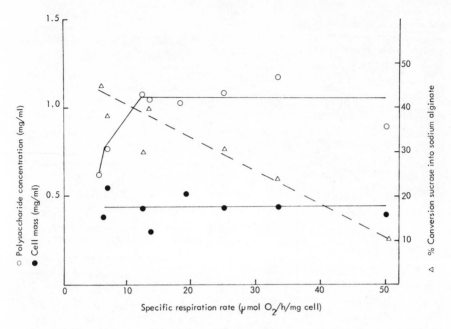

Figure 9. *Exopolysaccharide production by* Azotobacter vinelandii *at a range of res-piration rates*

Table 1.
Effect of growth-limiting nutrient on exopolysaccharide production by Azotobacter vinelandii

Growth-limiting nutrient	Cell Mass (mg/ml)	Specific Rate of polysaccharide production (mg/mg cell/h)
$MoO_4^=$	1.1	0.34
$PO_4^{\equiv}$	1.9	0.28
Fe^{++}	1.4	0.25
C(sucrose)	1.3	0.25
N_2	1.5	0.22
Ca^{++}	1.2	0.20
K^+	1.9	0.16
O_2	1.2	0.06

$$D = 0.15 \pm 0.01 \ h^{-1}$$

polysaccharide production was very much lower, with other conditions since under these conditions the cell mass was probably less active due to intracellular accumulation of poly-β-hydroxybutyrate (16). With the exception of O_2-limitation the specific rate of polysaccharide production varied by just a little over twofold, which considering the large changes in the physiology of the cell which are likely under the various limitations, is not very great. Some change was found however in the physical properties of the polysaccharide produced under the various limitations. Therefore although the specific rate of alginate production does not vary greatly with changes in fermentation conditions the yield of alginate in terms of the amount of sucrose utilised is mainly determined by oxygen availability and thus the respiration rate of the culture. Continuous culture studies have given us sufficient information on the control of alginate biosynthesis to choose conditions where improved yields of alginate can be obtained.

In summary, the rate of alginate synthesis per unit cell mass remains relatively constant over a range of conditions where the physiological state of the cell would be expected to vary widely, that is over a range of growth rates, over a range of respiration rates and with a variety of growth limiting nutrients. How this constant rate is obtained in terms of control mechanisms remains unclear. As yet we are unable to distinguish whether it is a relatively uncontrolled process or whether fine controls are necessary to obtain this constant rate. From these findings and our observations on other exopolysaccharide producing organisms, namely Xanthomonas campestris and a Pseudomonas sp. the ability to produce exopolysaccharide at similar rates under a variety of conditions could be much more general than has hitherto been recognised.

Literature Cited

(1)   Herbert, C., Ellsworth, R. and Telling, R.C. J. Gen. Microbiol. (1965), 14, 601-622.

(2)   Moraine, R.A. and Rogovin, P. Biotechnol. Bioeng. (1973), 14 225-237

(3)   Lawson, C.J. and Symes, K.C. Unpublished data.

(4)   Williams, A.C. Ph.D. Thesis, University College, Cardiff, U.K. (1974)

(5)   Neijssel, O.M. and Tempest, D.W. Arch. Microbiol. (1976), 107, 215-221

(6)   Silman, R.W. and Rogovin, P. Biotechnol. Bioeng. (1972), 14 23-31

(7)   Drummond, D.W., Hurst, E.L. and Percival, E. (1961). J. Chem. Soc., London, p. 1208-1216.

(8)   Larsen, B., Sandsrød, O., Haug, A. and Painter, T. Acta Chem.
            Scand. (1969), 23, 2375-2388.
(9)   Smidsrød, O. Disc. Faraday Soc. (1974),57, 263-274.
(10)   Gorin, P.A.J. and Spencer, J.F.t. Can. J. Chem. (1966) 44,
            993-998
(11)   Larsen, B. and Haug, A. Carbohyd. Res. (1971), 17, 287-296.
(12)   Haug, A. and Larsen B. Carbohyd. Res. (1971),17, 297-308.
(13)   Pindar, D.F. and Bucke, C. Biochem. J. (1975), 152, 617-622.
(14)   Still, G.C. and Wang, C.H. Arch. Biochem. Biophys. (1964)
            105, 126-132.
(15)   Downs, A.J. and Jones, C.W., FEBS Lett. (1975), 60, 42-46.
(16)   Dawes, E.A. and Senior, P.J. Adv. Microbiol. Physiol. (1973)
            10, 135-266.

Xanthan Gum from Acid Whey

MARVIN CHARLES and MOHAMMED K. RADJAI

Department of Chemical Engineering, Lehigh University, Bethlehem, PA 18015

Xanthan gum (from Xanthomonas campestris NRRL 1459A) has been produced from media containing deproteinized acid-set or culture-set cottage cheese wheys, the lactose contentsof which were hydrolyzed to glucose and galactose by means of immobilized lactase. Both glucose and galactose were used almost completely to give gum yields, productivities and final concentrations which were generally as good as, and in some cases better than, those obtained with comparable conventional media.With the exception of an anomalous pH history (the pH increased rather than decreased) when culture-set whey permeate was used, the fermentations followed courses typical of those previously reported. Details of media preparation, fermentation conditions, and experimental results will follow a brief discussion of cottage cheese whey and whey permeate.

Cottage Cheese Whey and Whey Permeate

Acid whey is the high BOD waste resulting from the manufacture of cottage cheese. Its composition (1) (see Table I) varies somewhat with the curd-setting process employed (and with milk composition, etc.) but in general it contains around 4% to 5% lactose, 0.8% to 1.0% protein (lactalbumin), and lesser quantities of acids, minerals, vitamins, etc. Most of the acid whey produced each year is run to waste resulting in considerable costs to dairies and communities. Furthermore, such disposal results in yearly losses of over 100 million lbs. of valuable and marketable whey protein (lactalbumin), which has excellent nutritional and functional properties, and over 500 million lbs. of lactose along with lesser but significant

quantities of organic acids and vitamins. Therefore
there is considerable economic incentive for the development of processes for direct utilization of acid
whey or for recovery and subsequent use of individual
acid whey components but the latter approach appears
to have greater potential in the forseeable future.

Table I. Acid Whey Composition (Typical)

	Culture Set	Acid Set
Lactose (wt %)	4.3-4.4	4.6-4.9
Protein (wt %)	0.8-1.0	0.9
Ash (wt %)	0.7-0.8	0.8-0.9
Lactic Acid (wt %)	0.6-0.8	--
Glucono-δ-Lactone (wt%)	--	0.04
Calcium (G/L)	1.2-1.3	1.3-1.4
Phosphorous (G/L)	0.7-0.8	1.9-2.1
Total Solids (wt %)	6.9-7.0	7.0-7.2
pH	4.3-4.7	4.1-4.5

Recovery of lactalbumin by the proven technology
of ultrafiltration offers considerable economic promise throughout most of the country and already has been
operated commercially. However, an important factor
influencing the economics of the recovery is the ulti-
mate use of whey permeate which is the by-product of
ultrafiltration and which contains a large quantity of
lactose, some low molecular weight protein, organic
acids, minerals, vitamins, and some other minor components. We require, then, economical uses for whey
permeate (2).
Many suggestions have been made for direct utilization of permeate including conversion to yeast and/
or alcohol (3). Fermentation technologies for both
are well known and it seems reasonable to expect that
there may be some cases in which such processes will be
economically feasible although it must be recognized
that the relatively low economic values of the products
might be a deterrent to investment. However, in the
absence of recent well-documented economic studies it
is difficult to make a satisfactory analysis particularly in light of the potential, but somewhat uncer-
tain, large-scale use of ethyl alcohol as a fuel.
Another approach involves the hydrolysis of whey
permeate lactose to glucose and galactose by means of
immobilized lactase (4,5,6,7). Widely discussed food
related applications of the "sweet permeate" so produced are based on the desire to recycle whey permeate
so as to eliminate disposal costs, to decrease sweet-
ener costs, and to circumvent nutritional and

functional problems associated with lactose. However, despite the fact that the hydrolysis can be performed for well under 10¢/lb of lactose (5), the "sweet permeate" may still meet with stiff competition from available corn and high fructose syrups since galactose is not as sweet as glucose and hence on a pound for pound (solids) basis the hydrolyzate is not as sweet as the already available syrups. Furthermore, it appears that demineralization will be required to make the hydrolyzate acceptable as a food ingredient and this will add considerably to its cost (5). These facts, coupled with the decline in sugar prices have cast some doubt on the very promising economic prognosis which existed for the use of "sweet permeate" as a food ingredient just a short time ago (6).

An alternative use of the hydrolyzate is as a fermentation medium. There are many organisms which will metabolize both glucose and galactose (but not lactose) to products considerably more valuable than yeast or alcohol and whose nitrogen requirements are satisfied partially or completely by the low molecular weight permeate proteins. This is particularly true in cases where production of large quantities of cell mass is not required or even particularly desirable (e.g., in production of xanthan). Furthermore, demineralization of the hydrolyzate is generally not required for this application. Thus, insofar as use as a fermentation medium is concerned, hydrolyzed permeate has the following advantages:

- carbohydrate cost competitive with glucose
- adequate nitrogen and other growth factors for many applications
- utilizes a high BOD waste stream
- enhances economics of whey protein recovery.

It should also be noted that even if condensation is required to facilitate transportation, the cost of hydrolyzate would still be competitive with commercial dextrose.

The microbial production of xanthan gum is a particular example of an already successful commercial fermentation which uses a conventional glucose-containing medium but which can be conducted as well or better with a hydrolyzed whey permeate medium.

The Fermentation Process

Medium Formulation. The medium can be produced from either culture-set or acid-set cottage cheese whey by means of the process illustrated in Figure 1:
(a) Whey is filtered through a hollow-fiber

ultrafilter having a molecular weight cut-
off of 50,000 (HF 26.5-45 - XM50 cartridge,
Romicon, Inc., Woburn, Mass.).

(b)   The permeate, which has a pH of 4.1-4.6, is
hydrolyzed in a pilot-plant fluidized-bed
reactor containing A.niger lactase (Lactase,
L.P., Wallerstein, Chicago, IL) immobilized
on alumina particles (6,7).

(c)   The hydrolyzed permeate is then sterilized
and supplemented with sterile K_2HPO_4 and
$MgSO_4 \cdot 7H_2O$ to yield a medium whose composition is given in Table II.

(d)   The pH of the medium is adjusted to 7.0.

Table II. Hydrolyzed Permeate Medium[a]
           (Full Strength-Culture Set)

Glucose (wt %)	2.05
Galactose	2.05
Lactose	0.30
$K_2 HPO_4$	0.50
$Mg SO_4 \cdot 7H_2O$	0.01
Protein (Lowry)	0.20
Whey Acid	0.70
pH	7.0

(a) Medium also contains whey ash, acids,
    vitamins, etc.

While either acid-set or culture-set whey may be used,
it is important to note that the two are not equivalent
as will be illustrated below.

In some cases we have used the media described as
is while in others they have been diluted to approximately half-strength. Furthermore, we occasionally
have added small quantities of supplemental nitrogen
in the form of enzymically-hydrolyzed lactalbumin
(Edamin, Sheffield Chemical, Union, NJ). This was
proven to be particularly valuable when acid-set whey
was used.

Sterilization. Hydrolyzed whey permeate is a
complex medium containing sugars and low molecular
weight protein along with acids and various minerals
and hence some caution is necessary during steam sterilization, particularly when an autoclave is used as
it was in our case. We found that if the permeate was
sterilized at its natural pH (4.1-4.6) there was observable browning but there was almost no loss of
nutrients and inhibitory products were not formed to
any appreciable extent. Indeed, medium steam

sterilized at the natural permeate pH behaved as well
as filter-sterilized medium. On the other hand, steam
sterilization at pH 6.0 or greater resulted in severe
browning, considerable precipitation, loss of nutrients,
apparent formation of relatively high levels of inhibitory compounds and a generally inferior medium.

Fermentation Conditions. Bench-scale fermentations were conducted in 7 liter aerated, non-baffled
fermentors equipped with three pitched-blade turbine
impellers having tank diameter to impeller diameter
ratios of 1.8 to 1.0. We found that the use of multiple large impellers and the intentional removal of
baffles resulted in better mixing, oxygen transfer,
and productivity when the fermentation broth became
viscous, particularly at xanthan concentrations
greater than 1% (8). The fermentors were also equipped
with automatic foam controllers, dissolved oxygen
monitors, and pH control systems which added either
4N KOH or gaseous NH$_3$.

The seed culture was developed as suggested by
Moraine and his coworkers (9,10,11) and a 5% (V/V) seed
was used to inoculate the main fermentors in all cases.
Temperature was always maintained at 28°C and pH at
7.0 except when the pH remained above 7 as was typically the case when culture-set whey was used.

Analytical

Glucose, galactose, and xanthan concentrations
were measured at regular intervals. Glucose was
determined by means of a glucose-oxidase impregnated
membrane and galactose by means of a galactose-oxidase
impregnated membrane. Both were used in conjunction
with a YSI Model 23A glucose analyzer (YSI Instruments,
Yellow Springs, Ohio). The lactose content of unhydro-
lyzed whey was usually determined by first completely
hydrolyzing it with excess A.niger lactase (Lactase LP,
Wallerstein, Chicago, IL) and then measuring the
resulting glucose or galactose. In some cases the
galactose oxidase membrane, which responds to lactose
to an extent of 10-15% of its response to galactose,
was used to determine whey permeate lactose directly.
The lactose concentration in whey permeate used for
fermentations was calculated from the hydrolyzate
glucose concentration (which is equal to the galactose
concentration prior to inoculation) and the known initial permeate lactose concentration. The lactose concentration remained essentially constant throughout
all the fermentations performed as it was not

metabolized by X.campestris under the conditions
employed.

Xanthan was determined by first filtering fermentation samples to remove all suspended solids,
precipitating the xanthan in the filtrate by addition
of KCl(2%) and methanol (50-60%) and finally determining the dry weight of the precipitated gum.

Fermentation Modes

Both batch and repeated-batch fermentations were
performed. In repeated-batch operation a given fermentation cycle was terminated when the galactose concentration dropped to approximately 0.1% or when the
xanthan production rate became impractically low. At
that time, approximately 85-90% of the fermentor contents were replaced with fresh medium and a new cycle
was initiated.

Results and Discussion

Glucose/Galactose Medium. Fermentations were conducted using media based on 50/50 mixtures of pure
glucose and galactose to provide base-line data free
of ambiguities that might arise as a result of the
complex nature of whey-based media. The history of a
typical fermentation is given in Figure 2 and the composition of the medium used in Table III.

Table III. Glucose-Galactose Medium

Glucose (wt %)	1.3
Galactose	1.3
Edamin	0.06
$K_2 HPO_4$	0.50
$Mg SO_4 \cdot 7H_2O$	0.01
pH	7.0

The most interesting point illustrated by these
results is the simultaneous use of both sugars. Although galactose was used less rapidly than glucose
there was clearly no diauxie. Furthermore, both
sugars were utilized for gum production. Otherwise
the course of the fermentation was typical of those
reported by Moraine and his coworkers (9,10,11). The
final gum concentration of approximately 2% which represented a 77% yield was achieved in about 50 hours.

Culture Set Whey: Batch Fermentation. The history
of a typical batch fermentation based on culture-set

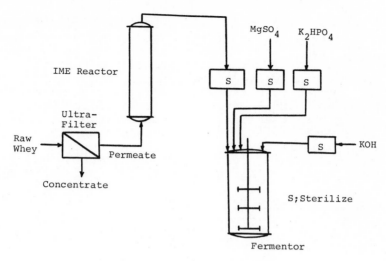

Figure 1.   Medium preparation

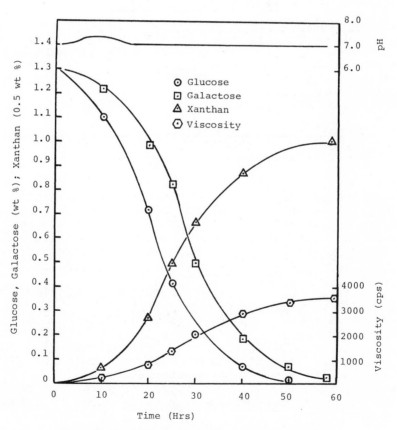

Figure 2.   Batch fermentation; glucose–galactose medium

whey permeate medium is illustrated in Figure 3 and
the composition of the medium used is given in Table I.
The most significant point to be noted here is that
the pH behavior was very different from that observed
by others using conventional glucose media or by our-
selves when we used the glucose/galactose medium. We
will return to this later. The other point worth
noting is that the final gum concentration of 3.5%
(in approximately 90 hours) represents an 85% yield
from the assimilable sugars which was considerably
greater than would have been expected on the basis of
previous reports. Again, we will return to this later.

        Acid-Set Whey: Batch Fermentation. Results of a
batch fermentation using half strength acid-set whey
medium supplemented with Edamin and having the composition given in Table IV are presented in Figure 4.

Table IV. Hydrolyzed-Permeate/Edamin Medium[a]
           (Half Strength-Acid Set)

Glucose (wt %)	1.3
Galactose	1.3
Lactose	0.2
Whey Protein (Lowry)	0.1
Edamin	0.06
$K_2 HPO_4$	0.25
$Mg\ SO_4 \cdot 7H_2O$	0.005
pH	7.0

(a) Medium also contains whey ash, acids,
    vitamins, etc.

In general, this history is the same as that for the
fermentation in which the glucose/galactose medium
was used although it did proceed somewhat more rapidly.
In particular, the pH behavior was typical and the
yield was within the range expected. It should also
be noted that media containing acid set whey but no
Edamin gave somewhat lower yields and longer fermentations.
        The reasons for the enhanced gum production and
anomalous pH behavior observed when culture-set whey
permeate was used are not clear. At this time we can
only speculate that differences in whey permeate com-
positions must be responsible, the primary differences
being in the concentrations of low molecular weight
whey protein, and in the concentrations and composi-
tions of the whey acid fractions. However, we can not
rule out other factors such as differences in vitamin
content.

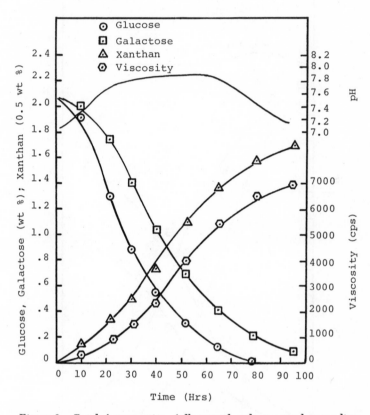

Figure 3.   Batch fermentation; full-strength culture-set whey medium

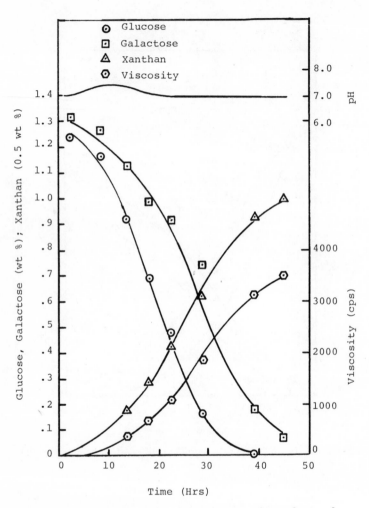

Figure 4.   Batch fermentation; half-strength acid-set whey medium

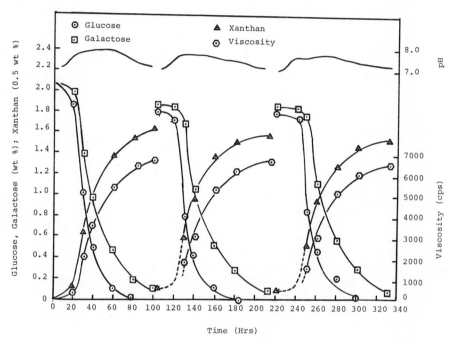

Figure 5. Repeated batch fermentation; full-strength culture-set whey medium

Repeated-Batch Fermentation. Results of a three-cycle repeated batch fermentation with full-strength culture-set whey medium (no Edamin) are illustrated in Figure 5. Other than the anomalous pH behavior and greater-than-expected yields the most notable feature of these results is that there was little change in fermentation history from cycle to cycle. However, it should be observed that there is a perceptible increase in lag time from one cycle to the next. At this time we can not say with certainty that this was actually a trend nor, if it was, can we predict the number of cycles which could be performed before the lag would become prohibitively long. However, we should note that because of the construction of the fermentor the culture retained as the seed for the next cycle always came from the very bottom of the vessel where mixing and aeration were particularly poor during the last hours of each cycle. This may have been the cause of the increased lag times.

Conclusion

Hydrolyzed whey permeate has been shown to be a suitable and competitive medium for the production of xanthan gum by X.campestris. It supports excellent yields and high final concentrations in both batch and repeated batch operation particularly when modified non-baffled agitation systems employing multiple large pitched-blade turbine impellers are used.

The authors wish to express their gratitude to the Pennsylvania Science and Engineering Foundation for supporting this work under PSEF Agreement #273 and to Romicon, Inc. for their generous gift of ultrafiltration cartridges used in this work.

Literature Cited

1. Personal communication, Lehigh Valley Dairy, Allentown, PA.
2. Melicouris, N., paper presented at Enzyme Technology Transfer and Utilization Conference, Lehigh University, Bethlehem, PA, May 27, 1976.
3. Goulet, J., paper presented at the First International Congress on Food and Engineering, Boston, Mass., August 10, 1976.

4. Coughlin, R. W., Charles, M., in "Enzyme Engin-
    eering" ed. Oye, E. K., and Wingard, L. B.,
    Plenum Press, N.Y., 1974.
5. Pitcher, W. H., in "Immobilized Enzymes for Indus-
    trial Reactors" ed. R. A. Messing, Academic Press,
    New York (1975).
6. Charles, M., Coughlin, R. W., paper presented at
    NSF/RANN Grantees Conference, University of Vir-
    ginia, Charlottesville, VA, May 19-21, 1976.
7. Charles, M., Coughlin, R. W., Allen, B. R., Paru-
    churi, E. K., Hasselberger, F. X., in "Immobilized
    Biochemicals and Affinity Chromatography", ed.
    Dunlay, R. B., Plenum Press, N.Y., 1974.
8. Charles, M., Zmuda, J., paper presented at AIChE
    Meeting, Nov. 28-Dec. 4, 1976, Chicago, IL.
9. Moraine, R. A., Rogovin, P., Biotech. Bioeng., $\underline{8}$,
    511 (1966).
10. Moraine, R. A., Rogovin, P., Biotech. Bioeng., $\underline{13}$,
    381 (1971).
11. Moraine, R. A., Rogovin, P., Biotech. Bioeng., $\underline{15}$,
    225 (1973).

4

Microbial Exopolysaccharide Synthesis

I. W. SUTHERLAND

Department of Microbiology, University of Edinburgh, West Mains Road,
Edinburgh, EH9 3JG, Scotland

The fate of a carbohydrate (or other) substrate supplied to
an exopolysaccharide-producing microbial cell depends on the
microbial species chosen.   As most results have been obtained
from bacterial species, this review will be concerned essentially
with the synthesis of exopolysaccharides by bacteria.

In some bacteria, given the correct substrate, exopolysaccharide may be formed without penetration of the cell membrane
by the substrate.   This is seen in dextran and levan-forming
cells supplied with sucrose or several of its analogues.
Examples are to be found in Leuconostoc mesenterioides, Streptococcus or Bacillus species.   Although this process has been
studied by various workers, (1,2) the polysaccharides formed are
more limited in their applications and current interest is centred
rather on species which form their polymer intracellularly then
excrete it into the medium.   The aim is therefore to consider a
series of processes by which substrates enter the microbial cells,
are modified by a series of enzymic processes and finally are
excreted in polymeric form from the microbial surface.   Much of
the information about these reactions has been gained from strains
producing polymers which have little or no commercial value, but
it is nevertheless possible to extrapolate many of the results and
thereby obtain a reasonable hypothesis for the mode of synthesis
of a polymer of given structure and to propose mechanisms for the
regulation of its biosynthesis.

Substrate Uptake   The substrate may enter the cell by one of
three mechanisms - facilitated diffusion, active transport or
group translocation.   The latter two processes, both of which are
endergonic, are of particular interest in the present context.
In active transport, the substrate enters the cell unaltered, but
the group translocation process involves the phosphorylation of
the substrate, the overall process being represented by:

$$X \ + \ PEP \longrightarrow X\text{-}P \ + \ pyruvate$$

The initial fate of the substrate is summarised in Fig.1.   In
Escherichia coli, the rate at which the bacteria grow on various
substrates is dependent on substrate uptake, irrespective of
whether active transport or group translocation systems are
involved (3).   Thus substrate uptake is one of the first limitations on exopolysaccharide production.   As yet, no attempts to
increase cell growth and hence exopolysaccharide production by
duplication of the genes concerned with active transport or with
group translocation appears to have been made.   In many bacteria,
this might not even be necessary, as several uptake mechanisms may
exist for each substrate i.e.   Although a specific substrate may
be transported by different mechanisms in different microorganisms
bacteria such as E. coli possess various mechanisms for uptake of
a single substrate such as galactose.   Differences can certainly
be expected between Gram positive and Gram negative bacteria or
between pseudomonads and enteric species.

The group translocation mechanisms involving phosphorylation
from PEP have been studied by Roseman and his colleagues (4) but
it is not clear whether the utilization of relatively large
amounts of PEP for substrate uptake lead to a reduction in the
amount of PEP available for other purposes.   If this does result
under conditions in which growth is limited by substrate uptake
and where high growth rates are used, the result might be a
reduction in the degree of pyruvylation observed in the polymer
excreted.

Intermediary Metabolism and Direction to Polymer Synthesis

Following the entry of the substrate into the cell and its
phosphorylation by either the group translocation mechanism or by
a hexokinase utilizing ATP, the substrate can be committed to
either anabolic processes or to microbial catabolism (Fig. 2).
If it suffers the latter fate, it is in effect wasted as far as
polymer production is concerned, although if it enters the TCA
cycle it may be converted to pyruvate or to acetate and thus
incorporated at a later stage into polymer.   The control of
catabolic processes will not be considered here.   The anabolic
fate of the substrate can still take one of several lines at this
stage.   If the microbial species under consideration is a Gram
negative species, forming exopolysaccharide, lipopolysaccharide
and glycogen, the carbohydrate may be converted to any one of
these.   In the proliferating bacterium, glycogen is rarely
synthesized, but its production is also differentiated from wall
polymer or extracellular polymer synthesis through the lack of
involvement of isoprenoid lipids.   The control of glycogen synthesis is exerted through allosteric regulation of ADP-glucose
synthesis (5), the first enzymic step in the pathway, which is
unique to glycogen synthesis (Fig. 3).   It may thus be worth
considering the isolation of ADP-glucose pyrophosphorylase mutants
if the bacterial strain in which we are interested produces large

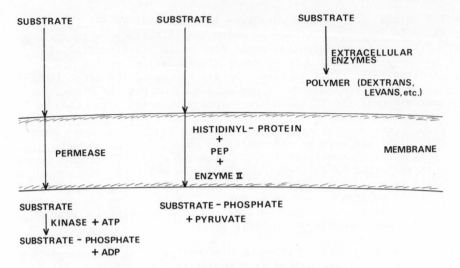

Figure 1.   Initial pathways for extracellular substrates

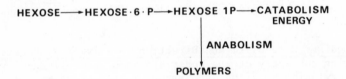

Figure 2.   Fate of hexose substrate

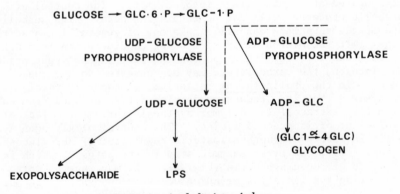

Figure 3.   Anabolic fate of glucose

amounts of glycogen and thus converts substrate to an unwanted product. This would eliminate the "drain" of glucose-1-phosphate into glycogen synthesis and away from the desired product. Such mutants would be particularly valuable if a two-stage production process was envisaged in which the second stage contained cells in an essentially non-proliferating environment, i.e. conditions under which large quantities of glycogen are normally synthesized. (Similar arguments would apply if the micro-organism produce poly-hydroxybutyric acid or trehalose rather than glycogen.)

The next precursor through which control can be exerted is the sugar nucleotide such as UDP-glucose. UDP-glucose pyrophos-phorylase is a key enzyme producing in many micro-organisms a precursor for both wall polymers and exopolysaccharide bio-synthesis. The level of UDP-glucose pyrophosphorylase synthesis appears to be almost unaltered in mutants defective in these polymers and this is reflected at least in the Enterobacteriaceae, in the level of UDP-glucose found in nucleotide pools of several strains (6). The strict control exerted by such enzymes as UDP-glucose pyrophosphorylase or TDP-glucose pyrophosphorylase (7) enables some micro-organisms to channel intermediates to one poly-mer or another. Thus, TDP-glucose is a precursor of TDP-rhamnose: for incorporation into one or more polymers. In species pos-sessing both enzymes mutual cross inhibition was observed, UDP-glucose inhibiting TDP-glucose pyrophosphorylase and TDP-glucose inhibiting UDP-glucose pyrophosphorylase (7). This could perhaps be predicted, as loss of synthesis of polysaccharide would lead to the accumulation of both glucose-containing sugar nucleotides. This double control is apparently restricted to micro-organisms in which polymers containing both sugars are found and is absent from micro-organisms lacking rhamnose-containing polysaccharides.

Similar control mechanisms are found in the formation of fucose as GDP-fucose from GDP-mannose. This was studied in bacterial species containing (i) D-mannose in their polysac-charides; (ii) containing L-fucose; and (iii) containing both D-mannose and L-fucose (8). In the first type, control of the rate of GDP-mannose synthesis occurred through GDP-mannose pyro-phosphorylase. In those bacteria in which GDP-mannose is solely a precursor in fucose synthesis, GDP-fucose controlled both GDP-mannose pyrophosphorylase and GDP-mannose hydrolyase through feed-back inhibition. When both mannose and fucose are present in polysaccharides produced by a single bacterium, each sugar nucleo-tide controlled its own synthesis (Fig. 4). Xanthomonas campestris is of particular interest because GDP-mannose and UDP-glucose most probably both serve as precursors for lipopolysac-charide and exopolysaccharide.

Further control of the nucleotide pool can occur through UDP-sugar hydrolases (9,10), although, as these enzymes in E. coli are

periplasmic, they may not necessarily have access to all the sugar
nucleotide formed by a cell but may be spatially separated from it
in normal cells. As several of the enzymes involved in sugar
nucleotide synthesis are membrane-bound, it is by no means clear
whether their products occur freely within the cytoplasm or
whether they are produced in close proximity to the enzymes which
require them for polymer synthesis.

    There is also the possibility of genetic regulation of
precursors specific to a particular polymer. The example of this
which has probably received most study, through the work of
Markovitz and his colleagues (11,12,13), is colanic acid synthesis
in certain bacteria of the Enterobacteriaceae. Knowledge of the
structure of colanic acid (14,15) (Fig. 5) reveals two monosaccharides, D-glucose and D-galactose, common to exopolysaccharide
and to wall polymers and two others, L-fucose and D-glucuronic
acid, unique to the polymer. Control of the exopolysaccharide
synthesis involved regulator genes; mutations in these genes led
to derepression and increased polysaccharide synthesis. As a
result of the derepression, increased production of the three
enzymes leading to GDP-fucose synthesis and under the control of
one regulator gene, was detected; increased formation of UDP-
glucose dehydrogenase (responsible for conversion of UDP-glucose
to UDP-glucuronic acid) occurred from mutation in another
regulator gene. As yet, the concept of such regulator genes as
those found in colanic acid formation, dominant on episomes but
recessive when located on the bacterial chromosome, is confined
to a few strains of E. coli, Salmonella etc. One should not
discount the possibility that polysaccharide production in other
genera and species is under similar genetic control, especially as
so little is known about the genetic systems of most exopolysaccharide-producing micro-organisms.

Formation of Exopolysaccharide   The construction of the repeating
units of the polymer is dependent on transfer of the appropriate
monosaccharides from sugar nucleotides to a carrier lipid –
isoprenoid alcohol phosphate. The sequence of reactions has been
well characterized through isolation of the products at each
transfer step (16) and through isolation and identification of
mutants (17) in two Enterobacter aerogenes systems. The series
of reactions for the strain studied by Troy et al. (16) was:

UDP-Gal   + P-lipid $\rightleftharpoons$ Gal-P-P-lipid   + UMP

Gal-P-P-lipid   + GDP-Man $\longrightarrow$ Man-Gal-P-P-lipid   + (GDP)

Man-Gal-P-P-lipid   + UDP-GlcA $\longrightarrow$ GlcA-Man-Gal-P-P-lipid   + (UDP)

GlcA-Man-Gal-P-P-lipd + UDP-Gal $\longrightarrow$ Gal-Man-Gal-P-P-lipid   + (UDP)
                                                     |
                                                   GlcA

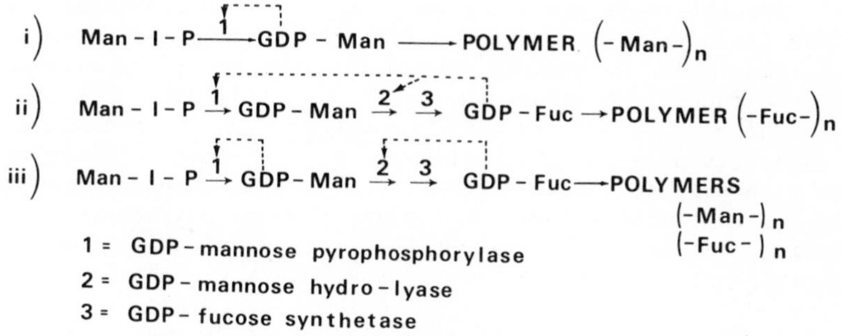

i) Man - I - P —1→ GDP - Man ——→ POLYMER (- Man -)n
ii) Man - I - P 1→ GDP - Man 2→ 3→ GDP - Fuc → POLYMER (-Fuc-)n
iii) Man - I - P 1→ GDP - Man 2→ 3→ GDP - Fuc → POLYMERS
(-Man-) n
(-Fuc-) n
1 = GDP - mannose pyrophosphorylase
2 = GDP - mannose hydro - lyase
3 = GDP - fucose synthetase

Figure 4. Control of mannose and fucose synthesis (after Kornfeld and Glaser, 1966)

PYRUV =4/6 β Gal
1
4
β GlcA
1
3
Gal
— 3 [β Glc 1 ——→ Fuc 1 ——→ 4 Fuc] 1 —
Ac

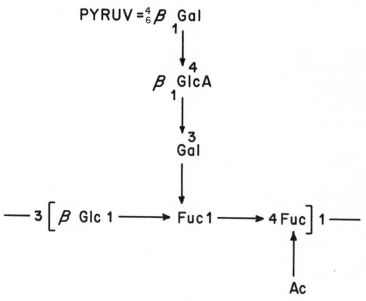

Figure 5

In the other strain studied (17), the first reaction also involved
transfer of a hexose-1-phosphate.   The methods employed in these
studies limited the size of fragment which could be identified as
being attached to the lipid.   The largest oligosaccharide characterized was an octasaccharide equivalent to two repeating units
(16).   The exact mechanisms involved in further chain elongation
and extrusion of exopolysaccharides is still unknown.   Recently,
two of the enzymes involved in E. aerogenes have been shown to be
extremely lipophilic proteins extractable from membrane preparations with acid butanol.   In this they resemble the isoprenoid
alcohol phosphokinase purified earlier from Staphylococcus aureus
(18) and a similar but not identical protein prepared from
E. aerogenes (19,20).

     The site of synthesis of lipopolysaccharide, which shares the
requirement for carrier lipid and also for certain of the sugar
nucleotides, has been identified as the cytoplasmic membrane (21).
Preliminary experiments in our labo·ratory have shown that it is
also the site of exopolysaccharide synthesis (Table 1).   Attempts
to purify the transferase enzymes by detergent solubilization
were unsuccessful; membrane proteins were solubilized but the
procedure usually led to partial or complete inactivation.

     Although studies of this kind have only been applied to a
limited number of micro-organisms, the general mechanisms appear
to be the same.   In the synthesis of the phosphorylated mannan
of Hansenula capsulata, both mannose and phosphate were derived
from GDP-mannose (22).   Although in this particular study there
was no attempt to demonstrate the involvement of lipid intermediates, they function in the formation of similar polymers in
microbial walls (23).   As the enzyme preparations used in these
studies were crude membranes, nothing is known about their
regulation, although in a series of non-polysaccharide-forming
E. aerogenes mutants, the amount of transferase activity appeared
to be lower than that found in wild type bacteria (17).

Isoprenoid Lipids in Exopolysaccharide Synthesis   The requirement
for isoprenoid lipids for exopolysaccharide synthesis is also
common to other repeating unit-containing glycan polymers located
external to the cell membrane i.e. the same carrier lipids are
used for synthesis of peptidoglycan, teichoic acids, lipopolysaccharide and exopolysaccharides.   Considerable indirect
evidence suggests that the availability of isoprenoid lipid
phosphate is one of the most critical factors affecting exopolysaccharide synthesis (24).   Any mutation affecting isoprenoid
lipid synthesis will thus affect exopolysaccharide production.
Various authors have indicated that bacteria contain 6.5-20 mg
isoprenoid lipid % dry weight (calculated from results in 25,26).
It has also been suggested that its availability could be controlled through phosphorylation of the free alcohol and dephosphory-

Table 1.   Location of Sugar Transferase Activities

E. aerogenes type 8, Glc-1-P and Gal I + II transferases assayed by standard techniques.

	Glc-1-P Transfer (%)	Gal Transfer (%)
Crude membrane	100*	100*
Spheroplast membrane	75	81
Cytoplasmic membrane	68	72
Outer membrane	3	0

*Activities were of the order of 0.154 nmol/mg protein/h and 0.282 nmol respectively.

lation of the alcohol phosphate and pyrophosphate (27). Unfortunately, Gram negative bacteria do not take up mevalonic acid and it is not possible to label the lipid precursors and thus obtain more accurate estimation of the amount present in cells than can be found from direct extraction. However, one possible way of increasing the isoprenoid lipid content appeared to be through selection for bacitracin resistance, since this antibiotic binds very strongly to isoprenoid lipids and effectively removes them from biosynthetic processess. Mutants with considerably elevated bacitracin resistance have been isolated in our laboratory and some undoubtedly yield more exopolysaccharide and show increased transfer of monsaccharides to lipid. (Other mutants were little different from wild type in all respects tested or had lost the ability to synthesize exopolysaccharide.)

It is also possible that some mutants defective in peptidoglycan synthesis might require less isoprenoid lipid than wild type cells, thus releasing more for exopolysaccharide synthesis. A mutant of this type has recently been isolated from E. coli B and, unlike the parent bacteria, produces exopolysaccharide (R.W. North, unpublished results). Similar observations have also been reported during attempts to prepare mutants for genetic engineering.

The reverse situation, reduced isoprenoid lipid content, is also difficult to study and can only be checked indirectly. Mutants with less lipid than wild type bacteria have not been characterized, but a group of CR (crenated) mutants isolated from E. aerogenes have characteristics which indicate that they may be conditional mutants of this type (28). These bacteria have rough colonial appearance at lowered incubation temperature and this has been ascribed to a reduced content of lipopolysaccharide. Exopolysaccharide is not synthesized until growth has ceased. The enzymes for polysaccharide synthesis are present in the bacteria grown at low temperature and on transfer to washed cell suspensions (non-proliferating conditions) exopolysaccharide is immediately formed in the presence or absence of chloramphenical. Thus no new enzymes have to be formed but at low temperature the synthesis of peptidoglycan - essential for cell viability - appears to take precedence over exopolysaccharide production and, to a lesser extent lipopolysaccharide synthesis. At 37°C, the mutants are identical in all respects tested to wild type bacteria. The mutants are not like classical membrane mutants, deficient in membrane phospholipid and susceptible to various detergents. Similar characteristics were observed in a polysaccharide-forming pseudomonad (29). The extracellular polymer was only produced late in the log phase of growth and in the stationary phase, having several of the attributes of a secondary metabolite. Could this too be due to insufficient isoprenoid lipid in the growing and peptidoglycan-forming bacteria?

In the literature, frequent reports of exopolysaccharide production being favoured by growth at low temperature are to be found; alternatively the polymer is said to be a product of the cells after growth has ceased. No satisfactory explanation for these observations has been provided, yet bacteria from the log or early stationary phases of growth appear to produce exopolysaccharide in washed suspension at similar rates. This could be due to limitation of exopolysaccharide synthesis during active growth through the availability of isoprenoid lipid; it would be needed for the formation of wall polymers until late in the log phase of growth. Limitation of carrier lipid also occurs in certain Salmonella mutants defective in lipopolysaccharide formation. Mutants forming the lipid-linked O-antigen but unable to transfer it to the appropriate acceptor have been characterized (30). Thus part of the normal isoprenoid lipid is no longer available for other processess, effectively reducing the total present in the bacteria. Mutants of this type could not produce exopolysaccharide although others defective in lipopolysaccharide synthesis but not accumulating lipid-linked glycans, had this capacity (31).

Several different types of mutations can thus affect isoprenoid lipid availability and consequently exopolysaccharide production. These are summarized in Fig. 6. The indirect evidence suggests a distinct series of priorities for isoprenoid lipid utilization. The essential wall polymer peptidoglycan has priority over lipopolysaccharide which in turn has priority over exopolysaccharide synthesis (Figs. 7 and 8). This could to some extent be achieved through spatial separation of the polysaccharide synthesizing systems within the microbial membrane but obviously requires further elucidation.

The final stages - modification and extrusion   As already discussed, the oligosaccharide repeating units accumulate on the carrier lipid and this type of mechanism probably applies to all exopolysaccharides other than dextrans, levans and related polymers (24). The mechanism could accommodate bacterial alginate synthesis if it is regarded initially as a homopolymer of D-mannuronic acid and is probably also valid for the glucans secreted by Agrobacterium species. However, many exopolysaccharides contain acyl and ketal substituents. Are these added while the repeating units are attached to lipid or at some later stage? (Fig.9). Preliminary evidence suggests that acylation occurs while the oligosaccharide is still attached to the lipid, but further studies are needed. This might indicate the lower degree of pyruvylation occurring in polysaccharide produced at higher growth rates (and higher resultant lipid turnover rates) reported in some species. The carbon source probably has no direct effect (Table 2). Considerable variations in acylation are found within a single polysaccharide. Thus, acetyl groups may occur on each repeating unit or on every second repeating unit in one E. aerogenes

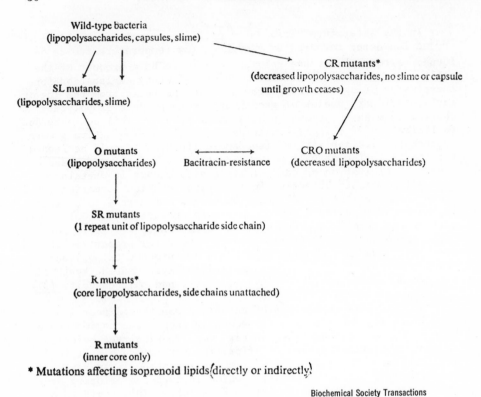

Wild-type bacteria
(lipopolysaccharides, capsules, slime)

SL mutants
(lipopolysaccharides, slime)

CR mutants*
(decreased lipopolysaccharides, no slime or capsule
until growth ceases)

O mutants
(lipopolysaccharides)   ←————→   CRO mutants
                        Bacitracin-resistance   (decreased lipopolysaccharides)

SR mutants
(1 repeat unit of lipopolysaccharide side chain)

R mutants*
(core lipopolysaccharides, side chains unattached)

R mutants
(inner core only)
* Mutations affecting isoprenoid lipids (directly or indirectly)

Biochemical Society Transactions

Figure 6.   How mutations affect the production of exopolysaccharides (31)

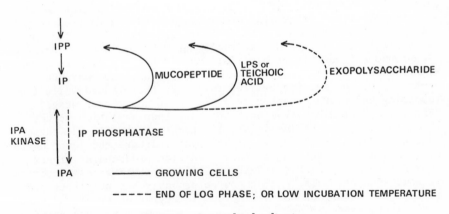

Figure 7.   Carrier lipid utilization

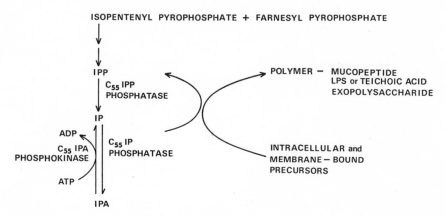

IPP – C$_{55}$ – ISOPRENYL PYROPHOSPHATE   IP – C$_{55}$ – ISOPRENYL PHOSPHATE
IPA – C$_{55}$ – ISOPRENOID ALCOHOL

ISOPENTENYL PYROPHOSPHATE + FARNESYL PYROPHOSPHATE

IPP
   C$_{55}$ IPP PHOSPHATASE
IP
ADP
C$_{55}$ IPA PHOSPHOKINASE   C$_{55}$ IP PHOSPHATASE
ATP
IPA

POLYMER – MUCOPEPTIDE LPS or TEICHOIC ACID EXOPOLYSACCHARIDE

INTRACELLULAR and MEMBRANE – BOUND PRECURSORS

Figure 8. Regulation of carrier lipids

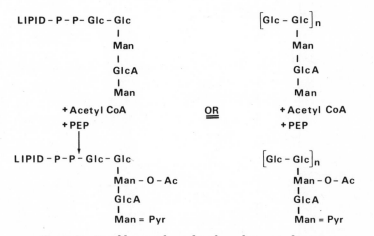

LIPID – P – P – Glc – Glc
   |
  Man
   |
  GlcA
   |
  Man
+Acetyl CoA
+PEP

OR

[Glc – Glc]$_n$
   |
  Man
   |
  GlcA
   |
  Man
+Acetyl CoA
+PEP

LIPID – P – P – Glc – Glc
   |
  Man – O – Ac
   |
  GlcA
   |
  Man = Pyr

[Glc – Glc]$_n$
   |
  Man – O – Ac
   |
  GlcA
   |
  Man = Pyr

Figure 9. Possible exopolysaccharide acylation mechanisms

Table 2. The Composition of a Pseudomonas Exopolysaccharide Derived from Growth on Various Substrates (Results of Williams, 1974)

Carbon Source	Hexose %	Deoxyhexose %	Acetate %	Pyruvic Acid %	Exopolysaccharide mg/ml
Glucose	75.0	15.2	3.52	5.55	3.23
Galactose	72.8	14.5	3.61	6.32	1.92
Mannose	74.0	18.5	3.67	5.48	1.74
Fructose	69.4	15.8	3.37	6.64	3.39
Xylose	67.3	15.8	3.84	6.46	2.20
Rhamnose	74.4	18.2	3.36	8.75	0.43
Arabinose	70.0	16.5	3.46	5.70	1.14
Ribose	72.1	17.9	3.16	5.36	1.89
Lactose	75.7	16.7	3.78	5.20	3.40
Maltose	79.7	17.4	4.03	6.28	2.65
Sucrose	73.6	16.9	3.75	5.81	2.87
Raffinose	74.0	18.0	3.90	6.96	0.60
Sodium pyruvate	79.4	17.1	3.46	6.10	1.55
Sodium succinate	78.8	18.5	3.52	6.22	1.30

strain (32). It has also been demonstrated that acetylation can be lost from a strain without loss of exopolysaccharide-synthesizing capacity (33). In contrast, loss of any enzyme contributing to the polysaccharide structure would lead to a non-mucoid variant. Similarly, pyruvylation also appears inessential for polysaccharide synthesis as, under certain growth conditions pyruvate groups can be lost but polysaccharide of apparently normal carbohydrate composition produced.

The exopolysaccharides studied so far, have mainly comprised repeating units with a single attached monosaccharide side-chain. It is possible that construction of the longer side chains found in xanthan gum or colanic acid might require some other mechanism such as construction of main-chain and side-chain units on separate carrier lipids prior to assembly. (An analogy can be found in lysogenic conversion, 34.)

The mode of final release from the isoprenoid lipid has not yet been demonstrated. It is unlikely that the process occurs through non-enzymic release of the increasingly hydrophilic elongating polysaccharide chain. This would probably leave the carrier lipid unavailable for further polysaccharide synthesis. In capsuleproducing strains, a ligase reaction may remove the polymer chain and attach it to the cell surface. It is unlikely that hydrolysis of the polysaccharide chain occurs at this stage unless a highly specific enzyme cleaves the terminal, phosphate linked monosaccharide:

$$\Downarrow$$
$$\text{Lipid} - P - P - \text{Glucose} - \text{Galactose etc.}$$

Enzymes reducing the degree of polymerization have been identified in alginate-producing bacteria (35) but the function of the enzyme is probably unconnected with polymer release of this type. Mutants unable to attach to the cell surface (S1 mutants) have been widely found, presumably through loss of the capsule attachment sites on the cell surface; other micro-organisms always produce exopolysaccharide as extracellular slime. The chain length of the polymer may also depend on the growth rate in a manner analogous to lipopolysaccharide side-chains (36), but this needs further study. Higher growth rate might lead to more rapid turnover of the carrier lipid and release of polymer of lower molecular weight. This is obviously important to the commercial producer. It may also be advantageous to use rough mutants (i.e. strains with surface defects) which autoagglutinate of flocculate and lead to easier polymer recovery. Thus exopolysaccharide production should be examined along with the synthesis of other polysaccharides and not in isolation.

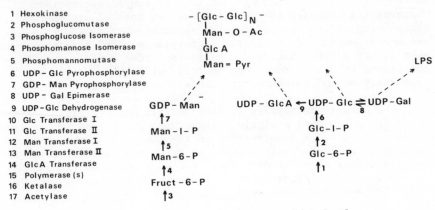

1 Hexokinase
2 Phosphoglucomutase
3 Phosphoglucose Isomerase
4 Phosphomannose Isomerase
5 Phosphomannomutase
6 UDP-Glc Pyrophosphorylase
7 GDP-Man Pyrophosphorylase
8 UDP-Gal Epimerase
9 UDP-Glc Dehydrogenase
10 Glc Transferase I
11 Glc Transferase II
12 Man Transferase I
13 Man Transferase II
14 GlcA Transferase
15 Polymerase (s)
16 Ketalase
17 Acetylase

Figure 10. Biosynthesis of Xanthomonas polysaccharides

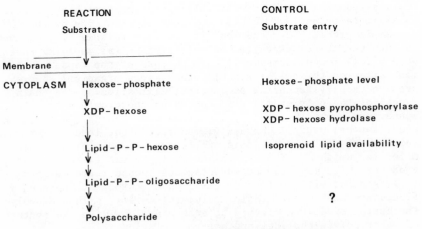

Figure 11. The control of polysaccharide synthesis

Summary

The production of microbial exopolysaccharides involves a relatively large number of enzymes, some of which are involved in the formation of other polysaccharides while others are unique to exopolysaccharide synthesis. By extrapolation from results obtained with other species, a biosynthetic pathway for X. campestris polysaccharide can be constructed (Fig. 10). Loss of most of these enzymes leads to loss of polysaccharide production, but variations in acylation or ketalation occur and may be of importance to the industrial microbiologist. Control of polysaccharide synthesis probably occurs at a number of levels (Fig. 11), and some mutations with altered polysaccharide regulation may have advantageous properties.

Literature Cited

1. Gibbons, R.J. and Nygaard, M. Arch. oral Biol., 13, 1249-1249 (1968).

2. Smith, E.E. FEBS Letters, 12, 33-37 (1970).

3. Herbert, D. and Kornberg, H.L. Biochem. J., 156, 477-480 (1976).

4. Roseman, S. In 'Metabolic Pathways' Ed. Hokin, L.E., 6, 41-89 (1972). Academic Press, London and New York.

5. Preiss, J. In 'Current Topics in Cellular Regulation' 1, pp. 125-160 (1969).

6. Grant, W.D., Sutherland, I.W. and Wilkinson, J.F. J. Bact., 103, 89-96 (1970).

7. Bernstein, R.L. and Robbins, P.W. J. Biol. Chem., 240, 391-397 (1965).

8. Kornfeld, R.H. and Ginsburg, V. Biochim. Biophys. Acta, 117, 79-87 (1966).

9. Ward, J.B. and Glaser, L. Biochem. Biophys. Res. Commun., 31, 671-6 (1968).

10. Ward, J.B. and Glaser, L. Arch. Biochem., 134, 612-622 (1969).

11. Lieberman, M.M. and Markovitz, A. J. Bact., 101, 965-972 (1970).

12. Lieberman, M.M., Shaparis, A. and Markovitz, A. J. Bact., 101, 959-964 (1970).

13. Markovitz, A. In 'Surface Carbohydrates of Prokaryotes', Ed. Sutherland, I.W., Academic Press, London and New York (In press).

14. Lawson, C.J., McCleary, C.W., Nakada, H.I., Rees, D.A., Sutherland, I.W. and Wilkinson, J.F. Biochem. J., 115, 947-958 (1969).

15. Sutherland, I.W. Biochem. J., 115, 935-945 (1969).

16. Troy, F.A., Frerman, F.A. and Heath, E.C. J. Biol. Chem., 246, 118-133 (1971).

17. Sutherland, I.W. and Norval, M. Biochem. J., 120, 567-576 (1970).

18. Sandermann, H. and Strominger, J.L. J. Biol. Chem., 247, 5123-5131 (1972).

19. Poxton, I.R., Lomax, J.A. and Sutherland, I.W. J. Gen. Microbiol., 84, 231-233 (1974).

20. Lomax, J.A., Poxton, I.R. and Sutherland, I.W. FEBS Letters 34, 232-234 (1973).

21. Osborn, M.J., Gander, J.E. and Parisi, E. J. Biol. Chem., 247, 3973-3986 (1972).

22. Mayer, R.M. Bio chim. Biophys. Acta, 252, 39-47 (1971).

23. Lennarz, W.J. and Scher, M.G. Biochim. Biophys Acta, 265, 417-441 (1972).

24. Sutherland, I.W. In "Surface Carbohydrates of Prokaryotes", pp. - , Academic Press, London and New York (In press).

25. Umbreit, J.N., Stone, K.J. and Strominger, J.L. J. Bacteriol. 112, 1302-1305 (1972).

26. Dankert, M., Wright, A., Kelley, W.S. and Robbins, P.W. Arch. Biochem., 116, 425-435 (1966).

27. Willoughby, E., Higashi, Y. and Strominger, J.L. J. Biol. Chem.,247, 5113-5115 (1972).

28. Norval, M. and Sutherland, I.W.   J. Gen. Microbiol., 57, 369-377 (1969)

29. Williams, A. Ph.D. Thesis, University College, Cardiff. (1974)

30. Kent, J.L.   and Osborn, M.J.   Biochemistry, 7, 4396-4408 (1968).

31. Sutherland, I.W.   Biochem. Soc. Trans., 3, 840-843 (1975).

32. Sutherland, I.W.   In "Surface Carbohydrates of Prokaryotes", pp.   -   , Academic Press, London and New York, (In press).

33. Garegg, P.J., Lindberg, B., Onn, T. and Holme, T.   Acta Chem. Scand., 25, 1185-1194 (1971).

34. Wright, A.   J. Bacteriol., 105, 927-936 (1971).

35. Madgwick, J., Haug, A. and Larsen, B.   Acta Chem. Scand., 27, 711-712 (1973).

36. Collins, F.M.   Aust. J. Exp. Biol. Med., 42, 255- 2 (1964).

5

Polysaccharide Formation by a *Methylomonas*

KAI T. TAM and R. K. FINN

School of Chemical Engineering, Cornell University, Ithaca, NY 14853

Extracellular microbial polysaccharides show great diversity as well as novelty in their structures and properties ($\underline{1}$). The applications of some of these biopolymers as stabilizers, emulsifiers, or thickeners in foods; as additives for recovery of petroleum by water flooding; as plasma extenders or as selective adsorbents in laboratory research, are well documented ($\underline{2},\underline{3},\underline{4}$). A new polysaccharide-producing bacterium called Methylomonas mucosa NRRL B-5696, was isolated from soil as an obligate methylotroph and the batch production of polymer and some of its properties have been described ($\underline{5},\underline{6}$). Kinetics for growth of the cells and for polymer production in shake flasks and chemostats are reported here.

Materials and Methods

The bacteria were maintained on agar plates with a 3% (v/v) methanol basal medium which contained 3.0 g KH_2PO_4, 3.7 g Na_2HPO_4, 2.5 g $NaNO_3$, 0.4 g $MgSO_4 \cdot 7\ H_2O$, 0.07 g Fe $(NH_4\ SO_4)_2$, 0.025 g Ca $(NO_3)_2 \cdot 4\ H_2O$, 0.001 g $ZnSO_4 \cdot H_2O$, in one liter of distilled water.

Methanol concentration was determined by a gas chromatograph with a flame ionization detector using ethanol as the internal standard. Cell dry weight was calibrated against a modified Lowry's protein assay ($\underline{7}$), and the latter was used for routine measurements. Polysaccharide concentration was expressed as glucose equivalent by the phenol-sulfuric acid method of Dubios et. al. ($\underline{8}$) with D-glucose as standard. Effluent gas composition was analyzed by a Fisher-Hamilton gas partitioner, model 29, using helium as a carrier gas. Dissolved oxygen measurements were made with membrane probes constructed as described by Johnson and Borkowski ($\underline{9}$, $\underline{10}$). Polymer was recovered by acetone precipitation ($\underline{11}$). Viscosity was measured in a Brookfield Synchro-Lectric Viscometer, model LVT with U.L. adaptor. Fermenter broths diluted in the range 1:5 to 1:10, were first degassed in vacuum, and then viscosity measurements for each dilution were made at various shearing rates. In some cases, viscosity of the

final broth was determined at a shearing rate of 30 RPM with a
No. 3 spindle using 150 ml of broth contained in a 200 ml beaker.
    Methanol is a toxic substrate for bacteria; even for metha-
nol utilizing organisms a concentration below 1.0% may inhibit
the growth of many strains (12,13). Therefore the effect of methanol concentration on the growth of M. mucosa was studied in
shake flasks. To do this, 250 ml portions of low phosphate medium (basal medium but with only half the amount of phosphate) in
1-liter indented flasks were inoculated with seed from a chemostat operating at a dilution rate of 0.25 hr^{-1} and at a steady-
state effluent methanol concentration of 1.0 v/v%. Methanol concentrations in the range 0.14 to 2.0% (v/v) were investigated.
Specific growth rates at 30°C and 350 RPM rotation of the shaker
incubator were determined in the time period when the maximum
change in the methanol concentration was less than 10% of the
initial value in the flask. The specific growth rate as a function of methanol concentration was then plotted.

Results and Discussion

    Kinetics of Growth. The exponential growth data from shake
flasks indicate that methanol concentrations above 1% v/v are
inhibitory (Figure 1). Furthermore, a Lineweaver-Burk plot shows
that at concentrations less than 1% the data fit a Monod model
for substrate-limited growth. The extrapolated maximal specific
growth rate, μ_m., from Figure 2 is 0.725 hr^{-1}, (equivalent to a
generation time of 0.956 hr). This is about 3 times higher than
the average value for most of the methanol utilizing bacteria
reported in the literature (11), and is about twice that of Pseudomonad C (14), the fastest growing methanol bacteria reported.
Such a fast growth rate makes M. mucosa attractive as another
bacterium for single-cell protein production. The high specific
growth rate observed in shake flasks was later confirmed by a
cell washout experiment in a chemostat, where the maximal specific growth rate was measured as 0.719 hr^{-1}.
    The other kinetic constant, K_S in the Monod model, was found
to be 0.20 M methanol. This value is two orders of magnitude
larger than the value of 0.00375M (120 mg/l) reported for Hansenula polymorpha - a thermophilic methanol-utilizing yeast whose
growth kinetics also fit the Monod model (15). Recent studies on
the growth of Candida boidinii, another methanol-utilizing yeast,
show a K_S value as high as 0.02M (16). No other literature values
of K_S for methanol-assimilating bacteria are available for comparison. The value of K_S obtained is also much higher than values
obtained for microbial growth on other carbon sources, which generally range from 1 to 50 mg/l (17). Since M. mucosa is subcul-
tured in 3% methanol-salts medium which is inhibitory for most
other methanol-assimilating bacteria, the bacterium must have
developed a transport mechanism that regulates a slow permeation
of substrate into the cell in order to reduce the inhibitory

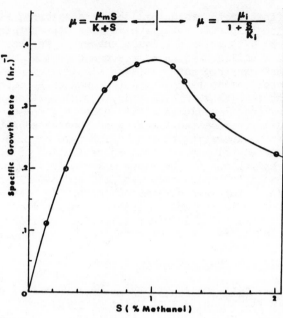

Figure 1. Specific growth rate at different initial substrate concentrations

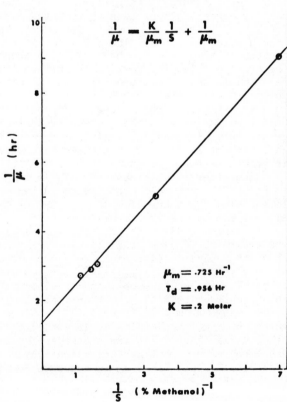

Figure 2. Lineweaver–Burk plot for the specific growth rate data

effect of the methanol. Also the polysaccharide slime is an additional barrier for the diffusion of methanol into the cell. The
high K_S implies that a low overall affinity for methanol should
be expected. The good agreement of the extrapolated μ_m with the
washout datum adds confidence to the accuracy of the kinetic constants. The implication of such a high value for K_S is that a
stable reactor can be operated at a doubling time as short as 1.8
hr for M. mucosa in a carbon-limited chemostat.
   Figure 3 shows that data for the substrate-inhibitory region
fit the model,

$$\mu = \frac{\mu_i}{1 + S/K_i} \qquad (1)$$

where
$$\mu_i = 6.05 \text{ hr}^{-1}$$

$$K_i = 18.4 \text{ m Molar}$$

The fact that data fit the two-parameter models does not tell us
the exact mechanism of inhibition or growth stimulation at the
molecular level. However, no further experiments were done to
elucidate the mechanism or site of inhibition because the prime
objective of determining the safe operation range for a carbon-
limited chemostat was obtained in this set of experiments.

   Respiration Kinetics. From the depletion rate of dissolved
oxygen and an average cell mass of 0.152 mg in the Yellow Springs
Dissolved Oxygen monitoring chambers, the specific respiration
rates were calculated for different initial methanol concentrations. In the absence of substrate inhibition, Michaelis-Menten
kinetics fit the respiration rate data as indicated by the
straight line in the Lineweaver-Burk plot (Figure 4). The cells
demonstrate a high affinity for methanol as suggested by the low
value of K_r, 8.1 μmolar methanol. The maximum respiration rate
(extrapolated) is 33 mMole O_2/(g cell, hr), which is slightly
higher than the average value of 26.6 mMole O_2/(g, hr) obtained
from an oxygen balance in the chemostat operating with a steady-
state effluent methanol concentration between 0.6% and 1.5% (v/v).
The lower chemostat value of V_m might be due to substrate inhibition. Compared with literature values (Table I), M. mucosa has a
K_r in the same order of magnitude as Hyphomicrobium. The maximal
specific respiration rate is about twice the highest rate listed
in the Table. A higher respiration rate is expected for M. mucosa
because of its very high specific growth rate. Another piece of
evidence that agrees with the extrapolated specific respiration
rate comes from separate batch experiments. At the point when
the dissolved oxygen reaches zero, the oxygen demand calculated
from the specific respiration rate should just equal the oxygen

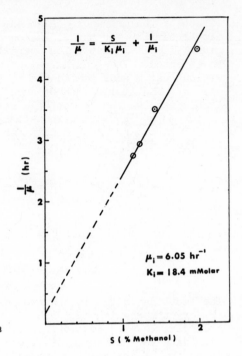

Figure 3. Kinetics of substrate inhibition for M. mucosa

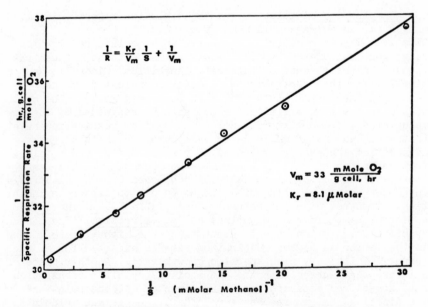

Figure 4.   Lineweaver–Burk plot for the respiration rate data of M. mucosa

supplied. The oxygen supply, based on gas chromatographic analysis of the influent and effluent gas, was 4640 mg O_2/hr, and the
predicted oxygen demand based on the above respirometer data (
33 mMole O_2 per g cell per hr) was 4780 mg O_2/hr.

The endogeneous respiration rate in methanol-free medium is
1.21 ± 0.05 mMole O_2/(g cell, hr) which agrees with the average
endogeneous rate of 1.3 ± 0.3 mMole O_2/(g cell, hr) obtained in
the Yellow Springs Dissolved-Oxygen Monitor Chamber, using cells
grown in different shake flasks.

Table I

Michaelis-Menten Kinetic Constants for the Respiration
of Bacteria Growing in Methanol

Reference	Organism	$K_r(\mu M)$	$V_m(\dfrac{mM\ O_2}{g,\ hr})$
Harrison (18)	Pseudomonas extorquens	20.4	10.5
Harrison (18)	methane utilizing Pseudomonad	50.0	4.15
Wilkinson (19)	Hyphomicrobium	8.33	0.0215
Wilkinson (19)	mixed culture	29400	0.024
This work	Methylomonas mucosa	8.1	33.0
Kim & Ryu (20)	Methylomonas sp.	-	18.0

Carbon-limited Chemostat. The straight line in the Lineweaver-Burk plot for the specific growth rate in the chemostat,
with methanol as the limiting substrate, suggests that Monod-type
growth kinetics fit the data (Figure 5).

$$\mu = \frac{\mu_m\ S}{K_s + S} \tag{2}$$

where
$$\mu_m = 1.43\ hr^{-1}$$

$$K_s = 0.583\ Molar$$

However, these constants do not agree with the shake flask data,
where $\mu_m = 0.725$ hr^{-1}, and $K_s = 0.20$ Molar. Such a discrepancy
is too large to be explained by the short-circuit of the flow or
other experimental errors. The only logical explanation is that
by continuously cultivating M. mucosa in the chemostat for over a
week, some variant was selected that had a lower affinity for
methanol and a faster growing rate.

When the specific methanol utilization rate (Q_m) is plotted
against dilution rate (D), a straight line is obtained (Figure 6).

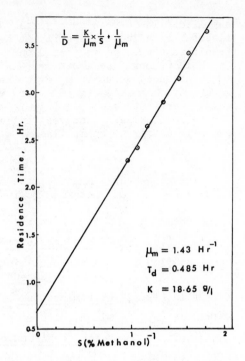

Figure 5.   Monod growth kinetics in carbon-limited chemostat

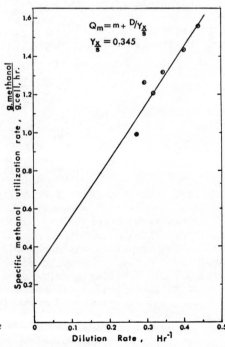

Figure 6.   Specific substrate utilization rate
correlation

This result confirms the validity of the empirical equation used by Pirt (17) and Nagai et al. (21):

$$Q_m = m + \frac{\mu}{Y} \tag{3}$$

For steady-state in a chemostat with no recycle, equation (3) becomes

$$Q_m = m + \frac{D}{Y_{x/s}} \tag{4}$$

where   m = maintenance coefficient for methanol
             = 0.26 g methanol/(g cell, hr)
$Y_{x/s}$ = 0.345 g cell/g methanol

The extrapolated m is the same order of magnitude as that reported for A. vinelandii (22), but an order of magnitude higher than that for other microorganisms (Table II). Unlike most of the microorganisms listed, (with the exception of A. vinelandii which forms poly-beta-hydroxybutyric acid) M. mucosa produces extracellular polysaccharides in addition to cell tissues and carbon dioxide. The formation of extra storage product or polymer requires more carbon uptake, and therefore a higher value of the cell maintenance coefficient should be expected for A. vinelandii and M. mucosa as indicated in Table 2. The experimental yield coefficient $Y_{x/s}$ = 0.345 g cell/g CH_3OH is quite reasonable, because Mateles et al. reported $Y_{x/s}$ = 0.31 for their polymer-producing Pseudomonad C in shake flasks (24) and $Y_{x/s}$ = 0.54 in a chemostat that favored cell production (14). The cell yields of other methanol-utilizing bacteria, with no polymer production, range from 0.2 to 0.4 (11).

Haggstrom (25) estimated that for his methanol-utilizing bacteria, the efficiency of transforming the carbon from methanol into the carbon in cells would be 41% ($C_{biomass}/C_{methanol}$). If we assume the composition of M. mucosa is $C_5H_8O_3N$ and calculate the efficiency of carbon transformation to biomass from the experimental yield $Y_{x/s}$ = 0.345, the efficiency is 42.7%, about the same as the number as obtained by Haggstrom. The empirical formula $C_5H_8O_3N$ is used instead of the formula of $C_4H_8O_2N$ based on Hamer and Johnson's data because the latter predicts 13.7% in the cell, whereas the nitrogen content of M. mucosa is 11.0 + 0.5% a value in closer agreement with the formula $C_5H_8O_3N$ (Table 3).

The average yield coefficients $Y_{p/s}$ = 0.175 and $Y_{CO_2/s}$ = 0.483 are constant within the range of dilution rates studied (Figure 7). If the hypothesis is correct, that the carbon in the methanol only turns into cells, polymer and carbon dioxide then a carbon balance based on the sum of the three yield coefficients

Table II

Growth Yield and Maintenance Coefficients for $\underline{M.\ mucosa}$ and Other Microorganisms

Organism	Limiting Factor	Substrate	m $\frac{\text{g sub.}}{\text{g cell, hr}}$	$Y_{x/s}$ $\frac{\text{g cell}}{\text{g sub.}}$	m_o $\frac{\text{g } O_2}{\text{g cell, hr}}$	$\frac{\text{g cell}}{\text{g } O_2}$
Aerobacter aerogenes (22)	glycerol	glycerol	0.08	0.56	0.10	0.94
Azotobacter vinelandii (22)	oxygen	glucose	0.15	0.26	0.18	0.41
Saccharomyces cerevisiae (22)	glucose	glucose	0.02	0.50	0.02	1.10
Methane mix bacteria (23)	methane	methane	0.12	0.7	0.06	0.38
Methylomonas mucosa	methanol	methanol	0.26	0.345	0.039	0.425

Table III

Elemental Analyses of Methane and Methanol Utilizing Bacteria

Substrate	C	N	H	O	Other Elements	Reference
methane	46.7	9.48	7.1	36.72	-	Vary & Johnson (26)
methane	50.1	11.7	7.14	29.44	1.62 P	Sheehan et al. (23)
methane	47.9	11.0	7.0	30.1	4.0 ash	Hamer et al. (27)
methanol	48.0	11.4	---	---	-	Haggstrom (25)
methanol	45.0	11.0	---	---	-	Harrison et al. (28)
average:	47.5	10.9	7.1	32.12	2.4	
methanol	46.2	10.8	6.15	36.9		assuming a formula of $C_5H_8O_3N$

should add up to unity. In order to do the carbon balance, the following two assumptions were made: 1) there is 46.2% carbon in the cells as indicated by the empirical formula $C_5H_8O_3N$ and 2) there is 40% carbon in the polymer. Since the polymer is a heteropolysaccharide, the general formula for carbohydrate CH_2O is a good approximation. The overall carbon balance comes out to be (Figure 7):

$$Y_{p/s} \frac{0.40}{0.375} + Y_{x/s} \frac{0.462}{0.375} + Y_{CO_2/s} \frac{0.2725}{0.375} = 0.965$$

Stripping of the volatile methanol or a trace amount of byproduct formation during fermentation, such as the yellow pigment, will account for the 3.5% discrepancy in the carbon balance. Thus the data are internally consistent.

To account for all the yield data, the following stoichiometric equation can be written (11)

$$22\ CH_3OH + 15.5\ O_2 + 2NO_3^- 2H^+ \longrightarrow 2\ C_5H_8O_3N + 4\ CH_2O + 33\ H_2O$$

This equation predicts $Y_{p/s} = 0.171$, $Y_{x/s} = 0.369$, $Y_{CO_2/s} = 0.50$. These numbers agree with the $Y_{p/s} = 0.175$, $Y_{x/s} = 0.345$, $Y_{CO_2/s} = 0.483$ obtained from the experiment.

The experimental polymer yield of 17.5% is too low for any practical polymer production process. However, previous shake flask experiments performed with nitrogen limitation suggested that the polymer yield could be improved at the expense of cell yield (11). The feasibility of a continuous polymer production scheme with nitrogen as the limiting substrate will be investigated in the following section.

Nitrogen-limited Chemostat. To achieve nitrogen-limited growth, 1 g/L $NaNO_3$ was used in the feed and the flow rate of medium was adjusted so that the cell density was 1.63 ± 0.03 g cell/L for all the dilution rates (0.14 to 0.32 hr^{-1}). The growth of the cells was not oxygen limited since the dissolved oxygen, D.O., was always more than that equivalent to 30% air saturation.

The specific methanol utilization rate, Q_m, remained constant instead of increasing linearly with dilution rate as was the case for carbon-limited growth (Figure 6). The average Q_m is 0.97 + 0.015 g methanol/(g cell,hr). Since the cell concentration and the Q_m were essentially constant for all the dilution rates investigated, an increase in residence time implied that more methanol would be converted into polymer. Thus the specific polymer production rate, Q_p, should increase linearly with residence time, and this is shown in Figure 8 where polymer formation is expressed both as glucose equivalent, Q_g, and also as dry weight, Q_p,. The data for Q_p are scattered because of errors in dried weight determinations. However, the slopes of Q_g and Q_p should be the same, and at zero residence time, no

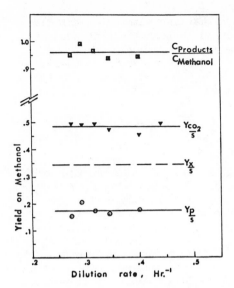

Figure 7. Yield coefficients for the carbon-limited chemostat

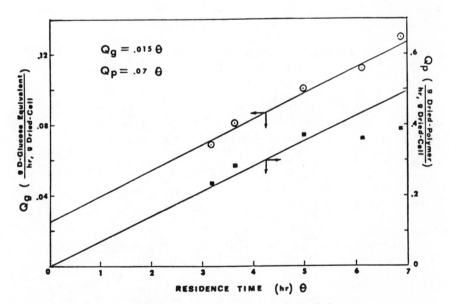

Figure 8. Polymer production in nitrogen-limited chemostat

polymer should be produced. A straight line with slope parallel
to Q_g and with zero intercept is drawn through the data for the
specific polymer production rate. For a given cell population
(X g/l), the total amount of polymer formed under nitrogen-limited conditions is given by:

$$X \int_{t_1}^{t_2} Q_p \, d0 = 0.035X \, (t_2^2 - t_1^2) \, (g \, polymer/l)$$

The yields of carbon dioxide based on methanol consumed did not
vary with dilution rate, D. The average value of $Y_{CO_2/s} = 0.47$
agrees well with the average value of 0.48 for the carbon-limiting case. However $Y_{p/s}$ and $Y_{x/s}$ varied inversely with each other
as dilution rate was changed (Figure 9). At D=0, the extrapolated
$Y_{x/s}$ is zero and $Y_{p/s}$ is 0.56, which suggests that the maximum
yield for the polymer is about 56% of the methanol consumed. This
apparent high projected yield might not be attainable in practice
because of dissolved oxygen limitation during the polymer formation phase. Even if the system were operated at half the maximum
yield, say at $Y_{p/s} = 0.28$, the performance would still be better
than for the carbon-limited case where $Y_{p/s} = 0.175$. The yield
data strongly suggest use of a nitrogen-limiting process for polymer production. A check for consistency of the data was made by
taking a carbon balance with the same assumptions as before, i.e.
46.2% carbon in cells and 40% carbon in polymer.

$$Y_{p/s}(1.058) + Y_{x/s}(1.232) + Y_{CO_2/s}(0.726) = C$$

The average value for C turned out to be 0.990 instead of 0.965
as in the carbon-limiting case. In other words, there was only
1% error in the carbon balance.
     When $Y_{p/s}$ is zero, the extrapolated maximum cell yield $Y_{x/s}$
is 0.555. Assuming that the carbon dioxide yield remains constant
at 0.47 in the absence of polymer formation, a carbon balance
gives a value of C as 1.02, i.e. 2% error in the carbon balance
when only cells and CO_2 are formed. The coincidence of the maximum values for the yield coefficients $Y_{x/s} = 0.56$ and $Y_{p/s} =
0.555$ suggests that the energy derived from catabolic processes
is used with approximately the same maximum efficiency for the
biosynthesis of either cells or polymer. In fact, these yield
data agree closely with the predictions based on "theoretical"
molar growth yields from ATP (29).
     From gas chromatographic analysis, the effluent air had an
average composition of 0.67 ± 0.02% carbon dioxide and 19.4 ±
0.15% oxygen. By an oxygen and carbon dioxide balance, the respiratory quotient (R.Q.) was found to be 0.418 mole CO_2/mole O_2.
The average specific oxygen consumption rate was 26.6 m mole

$O_2/(g$ cell, hr), which comes close to the extrapolated maximum value of 33 m mole $O_2/(g$ cell, hr) from the previous respiration study.

An interesting flocculation phenomenon was observed at the high dilution rates (F>1.2 l/hr). Cells tended to flocculate and settle much faster upon standing in a test tube at room temperature. However, after a shift to low dilution rate where more polymer was produced, the flocculating phenomenon disappeared. There are two possible explanations: either a mutant is formed or the flocculation is due to a concentration effect of the polymer. Only at a particular concentration of the anionic polymer that the interaction between the fixed amount of cell and the colloidal phosphate cation complex in the basal medium would bring the system to the isoelectric point and result in agglomeration and precipitation.

Agar plates inoculated with the precipitating cells gave the same type of colony as the normal cells. Also the rapid reversibility of the coagulating phenomenon achieved by changing the dilution rate (i.e. the amount of polymer formed) suggests that the second reason provides a better explanation.

Non-growth Associated Coefficient. It is apparent from the nitrogen-limited growth data that polymer production is non-growth associated. In order to test the extrapolated polymer yield data ($Y_{p/s}$ = 0.56) for a non-growth situation ($Y_{x/s}$ = 0) and to find coefficient for non-growth associated polymer production in the Luedeking (30) equation, $dP/dt = a\,dX/dt + bX$, a shake flask experiment was done using washed cells. Different amounts of washed cells suspended in phosphate buffer were used to inoculate nitrogen-free broth in indented flasks containing 1.29% methanol. The polymer production rates were linear for the first six to eight hours but decreased when the time of incubation increased beyond four generation times. The initial polymer production rate was plotted against the dried cell weight. A straight line was obtained as shown in Figure 10. The non-growth associated coefficient, b, obtained from the slope is 0.39 g polymer (g cell, hr). The average polymer yield for the five flasks was 0.59 ± 0.15 which agrees well with the extrapolated value of 0.56 from Figure 9. The relatively large error in the $Y_{p/s}$ calculation is due to the small quantities of methanol consumed in the first six hours; a difference of 0.01% methanol content would give 10% error in $Y_{p/s}$.

Nitrogen-limiting Batch. Based on the previous observations, a polymer production scheme with periodic nitrogen starvation was investigated. A basal medium containing 1.5 g/L $NaNO_3$ was used and two pulses of additional carbon and nitrogen (65 ml methanol and 2.15 g ammonium sulfate) were added at 9 1/2 hours and at 23 hours after the initiation of the batch run. For these experiments the Magnaferm fermentor had an aeration rate of 5 liters of

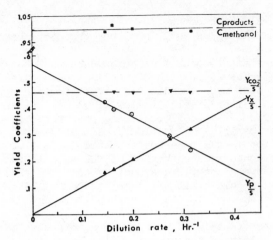

Figure 9. *Yield coefficients in nitrogen-limited chemostat*

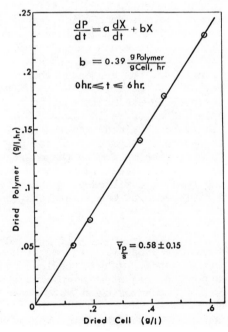

Figure 10. *Polymer formation rate in washed cell suspension*

air per minute, and a stirrer speed of 800 RPM.

For the exponential growth phase, the specific cell growth rate was 0.278 hr^{-1} (t$_d$ = 2.5 hr) which shifted to 0.102 hr^{-1} upon the addition of the first pulse of carbon and nitrogen. The low cell production rate was largely due to dissolved oxygen limitation, as shown in Figure 11. The concentration of dissolved oxygen remained zero after the consumption of 1% methanol.

The first-order rate constant for glucose production is 0.24 hr^{-1}. A lag of about two hours was observed before production of polymer resumed after the pulse addition of carbon and nitrogen. Since ammonium sulfate was used as a nitrogen source, no nitrite should accumulate to inhibit polymer production. Perhaps it takes time for M. mucosa to adjust to the concentration shock produced by a step increase of methanol to an inhibiting level. The important thing to note here is that the rate of polymer production does not decrease appreciably in a non-growth situation between the 14th and 23rd hours.

After the dissolved oxygen content reached zero, the mass transfer coefficient K$_L$a, as calculated from oxygen balance, was about constant. The average K$_L$a was 165 hr^{-1}. With the aeration and stirring parameters kept constant, the effect of the foam breaker on the oxygen mass transfer coefficient could be seen by a sudden decrease in K$_L$a from 165 to 120 hr^{-1} when the surface of the broth failed to reach the foam breaker (at the 34th hour). The termination of polymer production coincided with the drop in K$_L$a value. This observation suggests that mass transfer of dissolved oxygen may limit polymer production or else there is accumulation of toxic by-products in the batch.

Another interesting point was the continued methanol consumption at a linear rate of 0.332 g methanol/(l,hr) after both growth and polymer synthesis had stopped. This decrease in methanol could perhaps be accounted for by the cell maintenance requirement and the stripping loss. The final yield data at the end of the 56 hours were: Y$_{x/s}$ = 0.118, Y$_{p/s}$ = 0.408, and 1.897% solid produced for 4.55% methanol consumed. The maximum yield coefficients for polymer production in the batch should be calculated at the 38th hour when polymer production was terminated. Thus, disregarding the methanol loss in cell maintenance and stripping during the last 18 hours of incubation, the yield coefficients should be Y$_{x/s}$ = 0.122 and Y$_{p/s}$ = 0.452 while the total solid yield should be 0.574. The polymer yield approached the extrapolated maximum of 0.56 (Figure 8).

Semi-continuous Fermentation. An attempt to culture the bacteria continuously at low dilution rate was not very successful. The yield of polymer decreased after one week of continuous fermentation. Contamination at low steady-state methanol level and/or possibly culture degeneration became the major obstacles to successful operation at low dilution rate. Operation of a carbon-limited chemostat at a higher dilution rate had previously re-

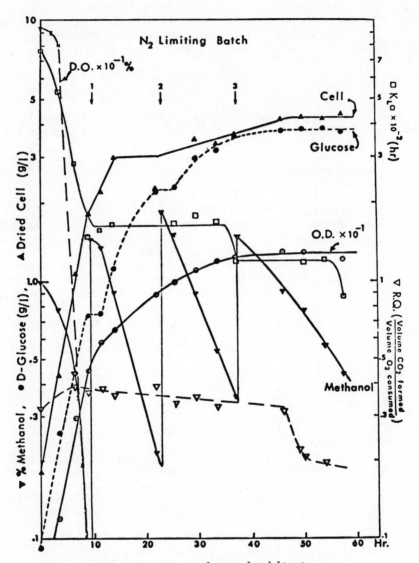

Figure 11.   Nitrogen-limiting batch kinetics

Table IV

Rate Constants for the Semi-continuous Operation in the 14 L Fermentor

	First Cycle		Second Cycle		
	Semi-continuous	Shake flask	Semi-continuous		Batch
Time, hr	0 to 32	33 to 47	3 to 47	30 to 40	42 to 64
$\dfrac{dX}{dt}\ \dfrac{g\ cell}{l,\ hr}$	.033	0	.014	.014	.044
$\dfrac{dP}{dt}\ \dfrac{g\ glucose}{l,\ hr}$	.065	.109	.057	.018	.031
$\dfrac{dS}{dt}\ \dfrac{\%\ CH_3OH}{hr}$	.088	.049	.054	.038	.045
RPM	500	350	500	500	550
$R.Q.\ \dfrac{mole\ CO_2}{mole\ O_2}$	.37 to .16	---	0.4 to .175	.175	.175

sulted in selecting a fast growing variant that was a poor polymer former. In order to avoid putting selective pressure on M. mucosa and to improve the process economics, a semi-continuous operation scheme was therefore considered.

The set-up consisted of two tanks in series; the first fermentor was the Magnaferm which served as a continuous cell propagator operating at relatively high dilution rate (0.42 1/hr). In this tank the steady state methanol concentration was kept above 1% so as to prevent growth of contaminants. Basal medium with 3% methanol and 2.5 g/l NaNO₃ was fed to the Magnaferm continuously. The effluent from the first tank was directed into the 14-1 fermentor where nitrogen-limited growth began. The nitrogen limitation not only favored polymer formation but also helped to prevent the growth of contaminants. After a fixed volume had accumulated in the second tank, effluent from the first tank was wasted(or diverted into another fermentor in actual plant operation) while the second tank was allowed to run as a batch process. A pulse of 1.7 g NaNO₃ and 84 ml methanol was added to the 14-liter fermentor when continuous feed stopped (Figure 12). At the 33rd hour of operation, the second tank was emptied and 250 ml of the broth was put in an indented flask and incubated at 350 RPM in a constant temperature (30°C) shaker. Then the second cycle of the continuous feed to the second tank began and no additional nitrate was added in this cycle.

The results of the two cycles of semi-continuous operation for the second tank are shown in Figures 12 and 13 and the rate constants are summarized in Table IV. In the first cycle, 2.776 g/l NaNO₃ was consumed in 33 hours and 4.22 g cell/l was produced. If all the available nitrogen had ended up in the cells, the percentage of nitrogen in the cells would be 10.82% which agrees with the value of 10.80% predicted by the empirical formula $C_5H_8O_3N$.

The measured cell and polymer production rates were all linear (Table 4). If the broth were allowed to incubate longer than 30 hours in the aerated tank, the rate of polymer production should slow down to about one-third of the initial value as suggested by data in the second cycle (Figure 12). However, the rate of polymer production actually increased from 0.065 to 0.109 g glucose/l, hr in the shake flask (at 350 RPM). This indicated that for the viscous broth, a shake flask had better aeration, and that a K_La value of 98 hr⁻¹ in the aerated tank was not high enough to meet the oxygen demand. Another piece of evidence for oxygen limitation was found in the second cycle of the operation in the second tank. A 10% increase in the agitation rate of the impeller, from 500 to 550 RPM, raised the polymer formation rate from 0.018 to 0.031 g glucose/(L, hr), and the cell production rate from 0.014 to 0.044 g cell/(L, hr). However, the K_La showed little observable change and remained constant at 120 hr⁻¹.

The respiratory quotient (R.Q.) is a very sensitive parameter that tells the age distribution of the population because the demand for oxygen and the evolution of carbon dioxide are not con-

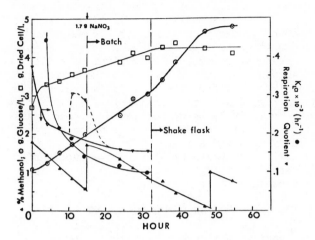

Figure 12.   First cycle semi-continuous fermentation

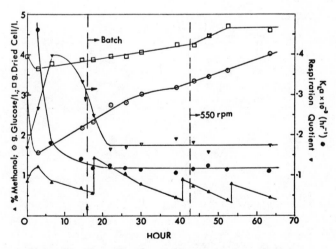

Figure 13.   Second cycle semi-continuous fermentation

stant during the different phases of the bacterial growth cycle. For example, in the first cycle of the semi-continuous operation, a pressure leak was developed in the first tank resulting in no overflow into the second tank and broth accumulation in the cell propagator. About an hour later (10-1/2 hour in Figure 12) the pressure was readjusted and about 0.6 liter of cells from the exponential growth phase was forced into the second tank. The R.Q. immediately jumped from 0.2 to 0.31 as indicated by the dashed line in Figure 12. The maximum in the R.Q. curve also indicated that a relatively large portion of exponentially growing cells was in the population during the continuous feeding stage in the second cycle (Figure 12).

Theoretically, the maximum respiratory quotient (R.Q. = 0.66) occurs when carbon in methanol serves only as an energy source and is completely oxidised to carbon dioxide and water.

$$2\ CH_3OH + 3\ O_2 \longrightarrow 4H_2O + 2\ CO_2 \quad 347\ Kcal/mole\ CH_3OH$$

However, when part of the carbon is diverted to cell and polymer synthesis, less carbon dioxide should be formed and the R.Q. should be less than 0.66. Since polymerization requires less energy than cell synthesis, the respiratory quotient should decrease monotonically as more polymer and fewer cells are formed. The experimental R.Q. data fall between 0.4 to 0.1 and the decrease in the respiratory quotient with the increase in the amount of polymer formed does in fact agree with the predicted general trend.

The final yield data for the semi-continuous experiment are summarized in Table V.

Table V

Final Yield Data for the Semi-continuous Experiment

Yield Constant	Tank 2	
	First Cycle	Second Cycle
$Y_{total\ solid/s}$	.32 (.387)	.34 (.424)
$Y_{x/s}$	.124 (.158)	.128 (.160)
$Y_{p/s}$	.196 (.229)	.212 (.264)

Numbers in the brackets show the yield constants, corrected for loss due to methanol stripping at a rate of 0.12 g Methanol/(1,hr) for 60 hours.

The yield data are lower than the best observed values of $Y_{x/s} = 0.122$, $Y_{p/s} = 0.452$ and $Y_{total\ solid/s} = 0.56$ in the nitrogen-limited batch (Figure 11). The poorer yield is due to the inferior oxygen transfer capacity of the second tank and, more importantly, to the nitrogen dosage scheme. In the nitrogen-limited batch culture, additional nitrogen source was added in a way that avoided prolonged periods of nitrogen exhaustion, and consequently there was a relatively large population of young cells in the broth. The fact that the R.Q. was between 0.4 and 0.33 in the first 50 hours of the batch operation as compared to the average R.Q. of less than 0.2 in the semi-continuous operation, supports the above argument.

In conclusion, a semi-continuous operation seems feasible because it is reproducible and because it minimizes problems of contamination or culture degeneration. The present operational procedure is not the optimal one. Improvements in aeration by installing a foam breaker and operating at a higher stirrer speed will help to bring the K_La value above $150 \ hr^{-1}$. Also the mode of nitrogen dosage can be revised so as to maintain a larger portion of young cells in the second fermentor (R.Q. between 0.3 to 0.4).

Literature Cited

1.   Bikales, N. M. (ed.) in "Water Soluble Polymers", pp 227-42, Plenum Publishing Corp., New York, N.Y., 1973.
2.   McNeely, W. H. in "Microbial Technology", H. Peppler (ed.), 381-402, Reinhold Publishing Corp., New York, N.Y., 1967.
3.   MacWilliams, D.C., Rogers, J. H., and West, T. J. in "Encyclopedia of Polymer Science and Technology", Vol. II, pp 105-126, Wiley-Interscience, New York, N.Y., 1973.
4.   Moraine, R. A. and Rogovin, P., Biotechnol. Bioeng. (1971), 13, 381-91.
5.   Tannahill, Alex L. and Finn, R. K., U.S. Patent 3,878,045, April 15, 1975.
6.   Finn, R. K., Tannahill, Alex L., and Laptewicz, J. E. Jr., U.S. Patent 3,923,782, Dec. 2, 1975.
7.   Herbert, D., Phipp, P. J., and Strange, R. E. in "Methods in Microbiology", Norris, J. R. and Ribbons, D. W. (eds.), Vol. 5B, pp 249-51, Academic Press, London, 1971.
8.   Dubios, M. et al., Anal. Chem. (1956), 28, 350-56.
9.   Johnson, M. J., Borkowski, J., and Engblom, C., Biotechnol. Bioeng. (1964), 6, 457-68.
10.   ibid. 9, 635-39.
11.   Tam, K. T., Ph.D. Thesis, Cornell University, Ithaca, N. Y., 1975.
12.   Van Dijken, J. P. and Harder, W., J. Gen. Microbiol. (1974), 84, 409-11.
13.   Whittenburg, R., Phillips, K. C., and Wilkinson, J. F., J. Gen. Microbiol. (1970), 61, 205-18.
14.   Battat, E., Goldberg, I., and Mateles, R. I., Appl. Microbiol. (1974), 28, 906-11.
15.   Levine, P. W. and Cooney, C. L., Appl. Microbiol. (1973), 26, 982-90.

16.   Pilat, P. and Prokop, A., Biotechnol. Bioeng. (1975), 17, 1717-28.
17.   Pirt, S. J., "Principles of Microbe and Cell Cultivation", 10-12, Blackwell Scientific Publications, Oxford, England, 1975.
18.   Harrison, D. E. F., J. Appl. Bacteriol. (1973), 36, 301-8.
19.   Wilkinson, T. G. and Harrison, D. E.F., J. Appl. Bacteriol. (1973), 309-13.
20.   Kim, J. H. and Ryu, D. Y., J. Fermentation Technol. (1976), 54, 427-36.
21.   Nagai, S., Mori, T., and Aiba, A., J. Appl. Chem. Biotechnol. (1973), 23, 540-62.
22.   Nagai, S. and Aiba, S., J. Gen. Microbiol. (1972), 73, 531.
23.   Sheehan, B. T. and Johnson, M. J., Appl. Microbiol. (1971), 21, 511-15.
24.   Mateles, R. I. and Chalfan, Y., Appl. Microbiol. (1972), 23, 135-40.
25.   Haggstrom, L., Biotechnol. Bioeng. (1969), 11, 1043-54.
26.   Vary, P. S. and Johnson, M. H., Appl. Microbiol. (1967), 15, 1473.
27.   Wilkinson, T. G., Topiwala, H. H., and Hamer, G., Biotechnol. Bioeng. (1974), 16, 41-59.
28.   Harrison, D. E. F., Topiwala, H. H., and Hamer, G., pp 491-5, "Fermentation Technology Today: Proc. IVth Int'l Ferm. Symp.", G. Terui (ed.), Soc. Ferm. Technol., Osaka, Japan, 1972.
29.   Abbott, B. J. and Gledhill, W. E., Adv. Appl. Microbiol. (1971), 14, 249-60.
30.   Luedeking, R. and Piret, E. L., J. Biochem. Microbiol. Technol. Eng. (1959), 1, 393.

Molecular Origin of Xanthan Solution Properties

E. R. MORRIS

Unilever Research, Colworth/Welwyn Laboratory,
Sharnbrook, Bedford. MK44 1LQ., Great Britain

The technological importance of Xanthan gum rests principally on the following unusual and distinctive rheological
properties in aqueous solution. (1,2,3)
1) Remarkable emulsion stabilising and particle suspending
    ability.
2) Extremely large shear dependence of viscosity, leading to
    pronounced thixotropy.
3) Little variation in viscosity with temperature under normal
    conditions of industrial utilisation.
4) High salt tolerance.
The aim of this paper is to provide a unified explanation of
the origin of these properties, at a molecular level.

Solution Viscosity.

Normally polyelectrolytes adopt a highly expanded conformation under conditions of low ionic strength, but collapse to a
more compact coil on addition of salt, due to charge screening.
Since polymer solution rheology is critically dependent on
molecular shape, these variations in coil dimensions are
normally reflected in large changes in solution viscosity (4).
Since the xanthan molecule is a polyanion, its maintenance of
viscosity with increasing ionic strength is therefore particularly
surprising, and indicates a considerable departure from normal
random coil behaviour. The temperature dependence of its
solution viscosity is also complex. In the presence of moderate
amounts of salt xanthan viscosity shows virtually no variation
with temperature, in contrast to the normal marked decrease in
polymer solution viscosity on heating. Under low ionic strength
conditions, such as exist when the polymer is dissolved in
distilled water, the temperature dependence of xanthan rheology
is even more unusual, showing an anomolous increase in solution
viscosity on heating, over a specific fairly narrow temperature
range (1), suggesting a sharp change in molecular conformation
over this range.

Optical Activity

Changes in polysaccharide conformation are frequently accompanied by large changes in optical activity (5-13), and in particular single wavelength optical rotation provides a sensitive and convenient index of chain conformation. We have therefore used this approach to investigate further the origin of this peculiar temperature profile (14,15). As shown in Figure 1 the anomolous viscosity behaviour coincides exactly with a large sigmoidal increase in optical rotation, such as has been shown to accompany order-disorder transitions in other polysaccharide systems (6-10). Indeed a simple quantitative relationship has been developed (5) to predict changes in optical rotation arising from changes in the dihedral angles between adjacent sugar residues in the polymer chain.
Interpretation of xanthan optical rotation is, however, complicated by the presence of acetate, pyruvate, and uronate groups, all of which absorb light at longer wavelengths than the polymer backbone and might therefore dominate optical rotation measurements in the visible region. To explore this possibility we have used circular dichroism to monitor directly the temperature dependence of the optical activity of these chromophores. As shown in Figure 2, there is a large negative shift in c.d. on heating. The observed optical rotation shift to less negative values at high temperatures is opposite in sense, and must therefore arise from changes in the far-ultraviolet where the electronic transitions of the polymer backbone are known to occur (16,17).

N.m.r. Relaxation

Chiroptical and rheological evidence therefore indicates that the xanthan molecule exists in solution at moderate temperatures in an ordered conformation which, under suitable conditions, can be melted out. To further test this conclusion we have used time-domain pulsed n.m.r. to probe molecular mobility. N.m.r. relaxation by energy transfer between adjacent nuclei provides a sensitive index of polymer flexibility, being extremely rapid for rigid molecules, but much slower for flexible coils, where thermal motions interfere with the exchange. At elevated temperatures, salt-free xanthan solutions show only the millisecond relaxation processes normal for disordered polysaccharides. At ambient temperatures, however, a much more rapid relaxation is observed in the microsecond range typical of rigid, ordered structures (18).
High resolution n.m.r. linewidth is inversely related to the rate of decay of magnetisation, and so freely moving molecules show sharp n.m.r. spectra, while for rigid polymers the linewidth is so great that the high resolution spectrum is so flattened

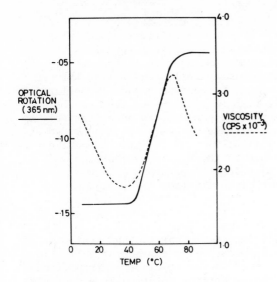

Figure 1. Comparison of optical rotation and viscosity variation with temperature for Xanthomonas *polysaccharide ("Keltrol")*

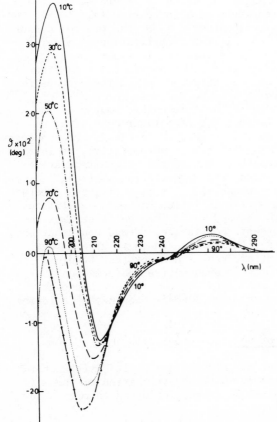

Figure 2. Keltrol. Effect of temperature on CD.

that no peaks are visible. Thus normal high-resolution n.m.r.
can be used to monitor order-disorder transitions (9-10). At
temperatures above the discontinuity in optical rotation and
solution viscosity, xanthan solutions show n.m.r. spectra typical
of a normal disordered polysaccharide coil. On cooling through
the transition region, however, the spectrum gradually collapses,
until finally no discernable high resolution spectrum can be
detected. This decay is conveniently monitored quantitatively by
measuring the area of the acetate and pyruvate resonances, which
occur as well resolved singlets at 2.1 and 1.5 ppm respectively
(Figure 3). As shown in Figure 4, the n.m.r. relaxation
behaviour follows the same sigmoidal temperature course as
optical activity and solution viscosity (see Figure 1).

The Ordered State

Rheological evidence (1) indicates that xanthan conformation
is critically dependent on the presence or absence of salt. To
investigate this we have followed the order-disorder transition
at various ionic strengths, using optical rotation as a convenient
index of conformational change. As shown in Figure 5, the transition shifts to higher temperature with increasing salt level,
until for ionic strengths above about 0.15 M, the ordered conformation persists up to 100°C. Similar stabilisation of ordered
structures by addition of salt is observed in other charged
polysaccharides (6-8), and is presumably due to the reduction of
electrostatic repulsions between neighbouring charged groups in
the compact, ordered state.

At constant ionic strength the temperature course of the
transition appears to be independent of polymer concentration
(Figure 6). This indicates that either the order-disorder process
is unimolecular, or that it is extremely co-operative, as in the
case of DNA (19). The breadth of the xanthan transition argues
against the latter explanation, and suggests intramolecular order.
The covalent structure of xanthan has only recently been determined (20,21), and consists of a cellulose backbone substituted
on alternate residues with charged trisaccharide sidechains, as
shown in Figure 7. We suggest that in the ordered conformation
the sidechains are aligned with the main chain to give a rigid
structure stabilised by intramolecular non-covalent bonding.
Definitive description of the ordered native conformation,
however, must await X-ray evidence. Such work is at present in
progress in Purdue University, and is described in the following
paper.

Molecular Interpretation of Solution Properties

Whatever the detail of the ordered state, its existence in
solution offers a satisfactory unifying explanation of the
unusual and valuable rheological properties of xanthan. In most

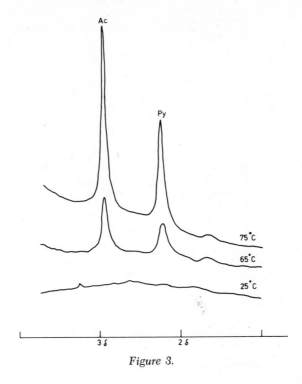

Figure 3.

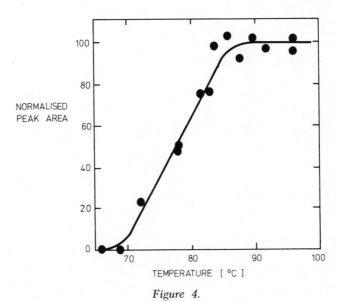

Figure 4.

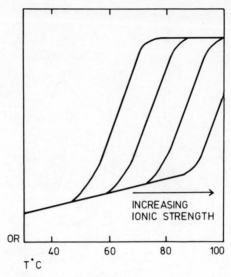

Figure 5.

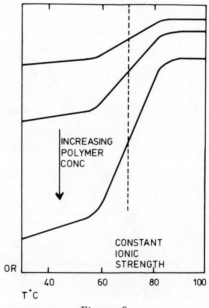

Figure 6.

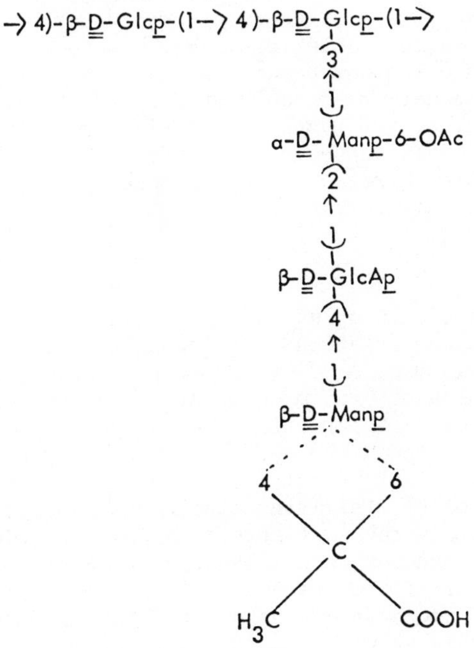

→ 4)-β-D̲-Glc$p$-(1→) 4)-β-D̲-Glc$p$-(1→)
                                 |
                                 3
                                 ↑
                                 1
                                 |
              α-D̲- Man$p$-6-OAc
                         |
                         2
                         ↑
                         1
                         |
              β-D̲-GlcAp
                      |
                      4
                      ↑
                      1
                      |
              β-D̲-Man$p$
              ⟋   ⟍
            4   6
              ⟍   ⟋
                 C
              ⟋   ⟍
        H$_3$C   COOH

Figure 7.

technologically important applications, sufficient salt is present to maintain the ordered conformation at all temperatures. The relative insensitivity of solution viscosity to addition of further salt is then a direct consequence of molecular rigidity.

The emulsion stabilising, particle suspending and thixotropic behaviour all point to the existence of appreciable intermolecular structure in xanthan solutions. Such an interpretation is entirely consistent with the known tendency of rod-like molecules in solution to align (22). Indeed, birefringence studies (2) give direct evidence of considerable molecular orientation in xanthan solutions. We therefore suggest that weak non-covalent associations between aligned molecules build up a tenuous gel-like network capable of supporting solid particles, liquid droplets, or air bubbles. Progressive breakdown of this network with increasing shear rate offers a direct explanation of the remarkable thixotropy which is perhaps the most valuable property of the polysaccharide.

Abstract

Xanthan exists in solution at moderate temperatures in a native, ordered conformation. At low salt levels this order may be melted out, as monitored by n.m.r. relaxation, optical rotation, circular dichroism, and intrinsic viscosity. We suggest that in the ordered conformation the charged trisaccharide sidechains fold back around the cellulose backbone, to give a rigid, rod-like structure. Increasing salt concentration stabilises this conformation by minimising electrostatic repulsions between the sidechains. At the salt levels encountered in most industrial situations, the ordered form is stable to above 100°C, hence the relative insensitivity of xanthan solution viscosity to temperature or further increase in ionic strength. Stacking of the rigid molecules in solution builds up a tenuous intermolecular network, giving rise to the other commercially attractive properties, such as suspending ability, emulsion stabilisation, and thixotropy.

Literature Cited

1.   Jeanes, A., Pittsley, J.E. & Senti, F.R. J. Applied Polymer Sci. (1961). 5, 519-526.
2.   Jeanes, A. In "Proceedings of the ACS Conference on Water Soluble Polymers" (Bikales, N.M., ed.), pp. 227-242, Plenum Press, New York, (1973).
3.   Glicksman, H. "Polysaccharide Gums in Food Technology", Academic Press, New York, (1970).
4.   Smidsrød, O. & Haug, A. Biopolymers. (1971). 10, 1213.
5.   Rees, D.A. J. Chem. Soc. (B). (1970). 877-884.
6.   Rees, D.A. & Scott, W.E. J. Chem. Soc. (B). (1971). 469-479.

7. Rees, D.A., Scott, W.E. & Williamson, F.B. Nature. (1970). 227, 390–393.
8. Rees, D.A., Steel, I.W. & Williamson, F.B. J. Polymer Sci. (C). (1969). 28, 261–276.
9. Bryce, T.A., McKinnon, A.A., Morris, E.R., Rees, D.A. & Thom, D. Faraday Discuss. Chem. Soc. (1974). 57, 221.
10. Dea, I.C.M., McKinnon, A.A. & Rees, D.A. J. Mol. Biol. (1972). 68, 153–172.
11. Grant, G.T., Morris, E.R., Rees, D.A., Smith, P.J.C. & Thom, D. Febs. Lett. (1973). 32, 195–197.
12. Morris, E.R., Rees, D.A. & Thom, D. J. Chem. Soc. Chem. Commun. (1973). p. 245.
13. Morris, E.R. & Sanderson, G.R. In "New Techniques in Biophysics and Cell Biology". John Wiley, London, (1972).
14. Morris, E.R., Rees, D.A. Young, G.A., Walkinshaw, M. & Darke, A. J. Mol. Biol. Submitted. (1976)
15. Rees, D.A. Biochem. J. (1972). 126, 257–273.
16. Balcerski, J.S., Pysh, E.S., Chen, G.C. & Yang, J.T. J. Am. Chem. Soc. (1975). 97, 6274–6275.
17. Pysh, E.S. Ann. Rev. Biophys. Bioeng. (1976). 5, 63–75.
18. Darke, A., Finer, E.G., Moorhouse, R. & Rees, D.A. J. Mol. Biol. (1975). 99, 477–486.
19. Zimm, B.H. J. Chem. Phys. (1960). 33, 1349–1356.
20. Jansson, P.E., Kenne, L. & Lindberg, B. Carbohyd. Res. (1975). 45, 275–282.
21. Melton, L.D., Mindt, L., Rees, D.A. & Sanderson, G.R. Carbohyd. Res. (1976). 46, 245–257.
22. Flory, P.J. Proc. Roy. Soc. ser. A. (1956). 234, 50–73.

7

Xanthan Gum—Molecular Conformation and Interactions

R. MOORHOUSE, M. D. WALKINSHAW, and S. ARNOTT

Department of Biological Sciences, Purdue University, West Lafayette, IN 47907

Xanthan Gum, the extracellular polysaccharide produced by the microorganism <u>Xanthomonas campestris</u> has found widespread industrial use (1,2,3) because of its unique rheological properties. The polysaccharide forms homogeneous aqueous dispersions and solutions exhibiting high viscosity, as well as having characteristics of both pseudoplastic and plastic polymer systems (4,5). Of particular significance is the atypical insensitivity of solution viscosity to salt effects and to heat, especially at high ionic strength. Molecular weight measurements (6) indicate polydisperse systems of high molecular weight (>2x10^6).

The primary structure of xanthan has recently been reinvestigated (7,8) and found to consist of pentasaccharide repeating units (I).

$$4)-\beta-\underline{D}-Glc\mathit{p}-(1{\to}4)-\beta-\underline{D}-Glc\mathit{p}-(1{\to}$$

$$\begin{array}{c} 3 \\ \uparrow \\ 1 \end{array}$$

(I)

$$\beta-\underline{D}-Man\mathit{p}-(1{\to}4)-\beta-\underline{D}-GlcA\mathit{p}-(1{\to}2)-\alpha-\underline{D}-Man\mathit{p}-6-OAc$$

$$\begin{array}{cc} 4 & 6 \\ & C \\ H_3C & COOH \end{array}$$

Pyruvate is attached on average to about one-half of the terminal mannose residues; O-acetyl groups correspond to one residue for each pentassaccharide repeating unit. When previously detected in bacterial polysaccharides, pyruvate has usually been observed on every repeating unit (9,10). However, the closely related polysaccharides from other <u>Xanthomonas</u> species (<u>11,12</u>) also show differing pyruvate contents (<u>13</u>).

We have prepared fibers of both xanthan and the related polysaccharide from <u>Xanthomonas phaseoli</u> (<u>14</u>). Using established techniques for fiber diffraction and computer aided model building (<u>15,16,17,18,19</u>) we have been able to examine the possible molecular conformations of xanthan. The almost indentical X-ray diffraction patterns, from a large number of polysaccharide samples from both <u>X. campestris</u> and <u>X. phaseoli</u>, indicates an overall similarity of molecular conformation and primary sequence.

Results and Discussion

It is usually possible to prepare specimens of long helical polymers in which the molecules are aligned with their long axes parallel. Often further organization occurs, but rarely to the degree of a three-dimensionally ordered single crystal. The xanthan X-ray diffraction pattern (Figure 1) showing both continuous intensity distribution and Bragg maxima, is characteristic of an ordered array of helices which have their axes parallel but are not further ordered (<u>20</u>). The presence of continuous diffraction along the layer lines indicates that the individual molecules have random translations along and rotations about their axes and are not packed into a well developed crystal lattice. However, destructive interference has occurred near the center of the equator, leaving one broad Bragg reflection of spacing 1.9 nm, the array of molecules therefore has some order when viewed down a molecular screw axis at sufficiently low resolution.

The layer line spacing is consistent with a helix of pitch 4.70 nm; the meridional reflections $(0,0,l)$ occurring only when $l=5n$, suggests a 5-fold helix. This gives a rise per <u>backbone</u> disaccharide of 0.94 nm (Figure 2). The steric effect of the branching mannose residue together with the consequent removal of the cellulose $O_{(3)A}$--$O_{(5)}$ hydrogen bond across alternate β-1,4 linkages (using the notation in Figure 2) means that the backbone can no longer have the 2_1 screw symmetry of cellulose. Instead of the usual extended β-1,4 ribbon (<u>21</u>), a more sinuous helix of the type shown in Figure 3 is obtained.

A priori we could have no preference for any of the four possible 5-fold helical models. The 5/1 and 5/4 conformations are right and left-handed respectively and have a single turn per helix pitch while the two other (5/2 and 5/3) models also differ by being right and left-handed and have two turns per helix pitch.

Initially molecular models for each of these four single helical possibilities, were examined assuming standard bond-lengths, bond-angles and sugar ring conformation angles (<u>15</u>). The models were further constrained to exhibit symmetry and periodicity consistent with the diffraction pattern. On the basis of a minimum steric compression comparison, the 5/1 (Figure 3) and 5/2 (Figure 4) right-handed helices were most favored,

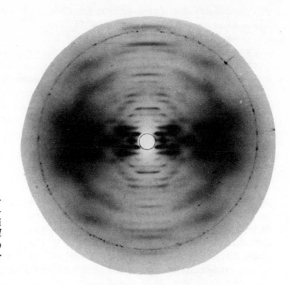

Figure 1. Diffraction pattern typical for both Xanthomonas campestris *and* Xanthomonas phaseoli *polysaccharides showing five-fold helical symmetry. The sharp Bragg reflection on the equator has a spacing of 1.9 nm.*

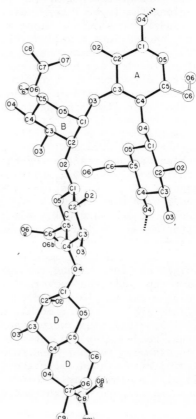

Figure 2. The pentasaccharide repeating unit of xanthan showing atom labeling and disaccharide backbone height. The unlettered residue and residue A are D-glucose, B and E are D-mannose, and C is D-glucuronate.

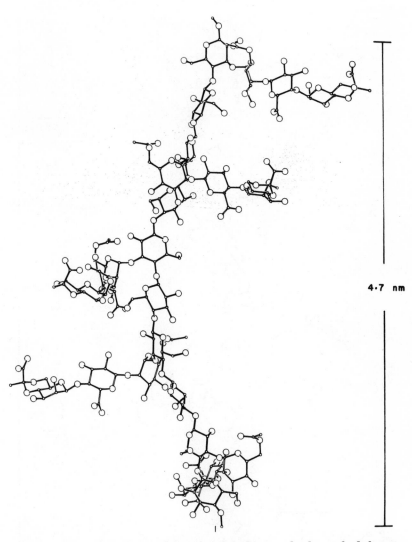

4·7 nm

Figure 3.   The isolated 5/1 xanthan helix viewed perpendicular to the helix axis

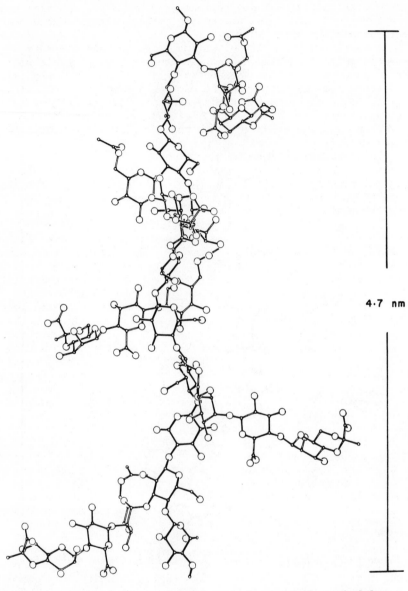

4·7 nm

Figure 4.   The isolated 5/2 xanthan helix viewed perpendicular to the helix axis

having glycosidic conformation angles within the normal oligo-saccharide ranges (Table 1) and no overshort non-bonded separa-tions. With the left-handed helices (5/3 and 5/4), minimization did not relieve all of the unacceptably short interatomic con-tracts even after optimization.

In isolation there is no driving force to hold the side chains close to the backbone and the isolated chain models sug-gest a diameter of 3.8 nm as opposed to a value of 1.9 nm obtained from lateral periodicities in the diffraction pattern. Studies on other branched polysaccharides favor the side chains lying roughly parallel to the backbone (18), and we have therefore undertaken a second study in which both packing and conformational variations were considered for each of the models.

The most symmetrical and commonly observed close packing of polymeric molecules, having nearly circular cross section, is hexagonal packing in which each molecule is equidistant from its 6 nearest neighbors but not necessarily further related. We therefore placed one xanthan helical chain in a hexagonal unit cell of side $\underline{a}$ = 2.19nm, $\underline{c}$ = 4.70nm, that is consistant with the equatorial Bragg reflection indexed as (100). Minimizing steric repulsion in this environment causes the side chain to fold down against the backbone.

Stereochemically both the 5/4 and 5/3 helices are unlikely as an unacceptable number of intramolecular overshort contacts persist after refinement. This reinforces our previous conclu-sion of right-handedness for the isolated chains. Although the 5/1 and 5/2 helices are sterically acceptable, the 5/1 exhibits the more favorable comparison with oligosaccharide conformation angles. It is of interest to note that the backbone conformation angles shown in Table I have varied little during the process of wrapping the side chains around the backbone. Further, the 5/1 'packed' helix (Figure 5) shows a number of potential intramole-cular hydrogen bonds (Table II and Figure 6). Relaxing the attractive interaction (hydrogen bond) terms in the refinement did not alter the molecular conformation. Only the additional influence of small perturbations to the conformation angles about the branching mannose linkage caused the stabilising influence of the hydrogen bonds to be lost.

The 'packed' 5/2 helix presents a much tighter structure than the 5/1 model while exhibiting some overshort intramolecular contacts and few potential hydrogen bonds and was considered unlikely on the basis of this analysis.

Our reasoning so far has been based on the premise that the equatorial Bragg reflection on the diffraction pattern (Figure 1), arises from the packing of single molecular entities, the pattern does not tell us what form these take. In our examination of inter-chain interactions, we have thus considered those inter-actions that can arise from some side-by-side arrangement of the 5/1 helices and also the case of coaxial multiple helices.

TABLE I

Comparison of backbone conformation angles in the isolated
and 'packed' 5/1 and 5/2 helical models

Angle	Range	Isolated helices 5/1	5/2	'Packed' helices 5/1	5/2
(a)	$-100 \to -161$	-136	-121	-148	-119
(b)	$-78 \to -98$	-63	-64	-76	-30
(c)	$-100 \to -161$	-111	-99	-98	-97
(d)	$-78 \to -98$	-92	-22	-81	-61

(a) $-\theta[C_{(1)A}, O_{(4)}, C_{(4)}, C_{(5)}]$

(b) $-\theta[O_{(5)A}, C_{(1)A}, O_{(4)}, C_{(4)}]$

(c) $-\theta[C_{(1)}, O_{(4)A}, C_{(4)A}, C_{(5)A}]$

(d) $-\theta[O_{(5)}, C_{(1)}, O_{(4)A}, C_{(4)A}]$

Using atom notation in Figure 2.

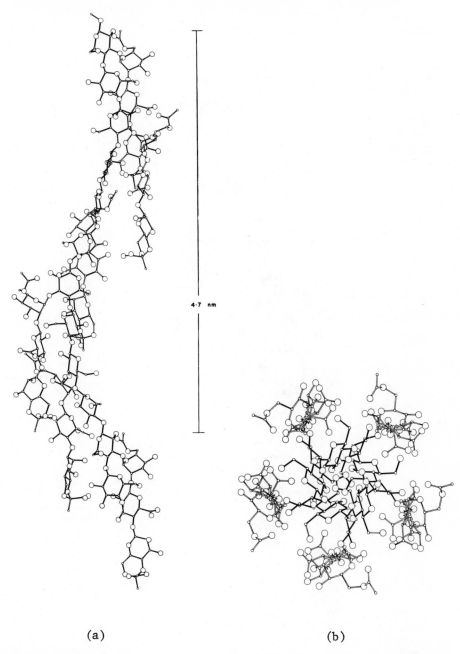

4·7 nm

(a)   (b)

Figure 5.   *The 'packed' 5/1 helix viewed (a) perpendicular to and (b) down the helix axis*

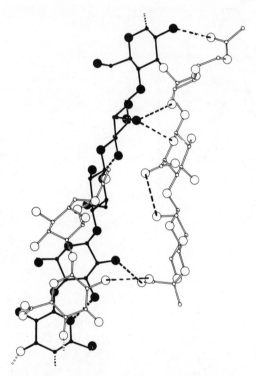

Figure 6. Possible hydrogen bonds (– – –) that may stabilize the molecule. Some adjoining residues are omitted for clarity, the backbone having solid bonds. See also Table II.

TABLE II

Possible attractive interaction in the $X,$ *campestris* 5/1 and 5/2 helices

Model	Overshort contacts (nm)	Potential Hydrogen bonds
5/1	none	$O_{(3)} \dashrightarrow O_{(5)A}$
		$O_{(2)} \dashrightarrow O_{(8a)}{}^D$
		$O_{(6)} \dashrightarrow O_{(5)C}$
		$O_{(2)A} \dashrightarrow O_{(7)B}$
		$O_{(3)B} \dashrightarrow O_{(6)}$
		$[\text{or } O_{(3)B} \dashrightarrow O_{(5)C}]$
		$O_{(2)D} \dashrightarrow O_{(6b)C}$
5/2	$O_{(5)A} \cdots H_{(4)A}$ (0.195 nm)	$O_{(3)} \dashrightarrow O_{(5)A}$
		$O_{(6)A} \dashrightarrow O_{(5)}$
		$O_{(3)B} \dashrightarrow O_{(5)C}$
		$O_{(3)C} \dashrightarrow O_{(5)D}$

Placing a single 5/1 helix in our hexagonal cell reveals few interactions with its nearest neighbors. This suggests that the helices are slotting into some groove that is wide enough to accomodate them without steric clashes. Alternatively the molecule could be considered as a rigid rod of polysaccharide surrounded by a cylinder of water, in which case very few polysaccharide-polysaccharide interactions would be apparent. Furthermore, as such a situation closely mimics the solution state, the unusual solution properties would probably arise from interactions between regions of 'ordered' water some of which may be tightly bound to the polysaccharide. Current X-ray fiber diffraction technology cannot enable us to locate this amount of water (18), possibly NMR studies on solutions may be able to locate 'ordered' water but without the detail that is sometimes available from diffraction studies (16,17). On drying the specimen for prolonged periods we note a reduction of over 50% in the cell volume consistent with a shrinkage in the Bragg spacing on the equator while the fiber axis dimension remains unaltered. Apparently the molecular conformation of the xanthan molecule survives drying with little change and a substantial quantity of water which fills out the structure is not firmly bound.

While it is possible to construct a double helical model, using the 5/1 single helix as precursor, in which the second coaxial strand is parallel to, and related to, the first by 180° rotation some apparently unresolvable overshort inter-strand contacts exist. It is possible that relaxing the summetry so that the parallel coaxial strands are not related by a 180° fiber axis rotation or are antiparallel to one another, could result in acceptable interactions between chains. Should this be the case it will still be necessary to obtain supporting evidence from other sources to demonstrate the existence of double helices. Normally this would take the form a comparison of the model with the X-ray intensity data from a crystalline diffraction pattern (e.g. 16,17,18) plus evidence from solution studies of bi-molecularity (e.g. 22). We would stress however that there is no evidence of double helices either in solution (27) or the solid state.

Recently, we have been able to obtain a diffraction pattern that exhibits increased crystallinity and which has been tentatively indexed on a tetragonal cell in which four 5/1 single helices will pack with the minimum of steric compression. A refinement using both stereochemical and X-ray intensity data has not yet been completed.

Conclusions

This preliminary study shows that the ordered conformation of xanthan in the condensed state, and probably in solution, is related to the 5/1 helix outlined here.

The interactions that occur in solution, giving rise to viscosity effects showing the characteristic of both flexible and stiff cross-linked regions (4,5) must arise from associations of the ordered 5/1 helical regions. The order/disorder transition seen with change of temperature in solution (23,24), would seem likely to arise from conformational changes primarily within the side chain as it moves away from its close association with the ordered backbone either accompanied by, or before, conformational changes in the backbone. This spreading of the 'arms' of the polysaccharide would cause an increased hydrodynamic volume and hence provide the viscosity stability noted at elevated temperatures (1,2,3,4).

Association in solution of single helices does not require gel formation, a fact that points strongly in favor of single helical xanthan, which does not show gelation at room temperatures. Weak gelation observed at temperatures close to 0°C is probably due to an aggregation phenomenon.

It is interesting to note that the 5/1 helix presents two distinct faces; one having the side chains and charged groups, the other essentially the cellulose backbone. As xanthan interacts synergistically with the β-1,4 linked galactomannans locust bean and guar gums, it is possible that this takes place at the cellulose 'groove' i.e. between similar β-1,4 linked glycans. It is thought that 'smooth' unsubstituted regions of the galactomannan are involved in the association (23,25).

More detailed interpretations of this continuing work will be published elsewhere (26).

Acknowledgements

We wish to thank Drs. A. Jeanes and P.A. Sanford, U.S.D.A., Peoria and Dr. I.W. Cottrell, Kelco, San Diego, for their generous gifts of samples.

Literature Cited

1. Jeanes, A. (1973) In "proceedings of the ACS Conference on Water Soluble Polymers", ed. N.M. Bikales, Plenum Press, New York. pp. 227-242.
2. Jeanes, A. (1974) J. Polymer Sci., Symp. No. 45, 209-227.
3. McNeely, W.H. and Kang, K.S. (1973) In "Industrial Gums" R.L. Whistler and J.N. BeMiller eds., pp. 473-497, Academic Press, New York.
4. Jeanes, A., Pittsley, J.E. and Senti, F.R. (1961) J. Appl. Polymer Sci., 5,519-526.
5. Jeanes, A. and Pittsley, J.E. (1973) J. Appl. Polymer Sci., 17,1621-1624.
6. Dintzis, F.R., Babcock, G.E. and Tobin, R. (1970) Carbohyd. Res. 13,257-267.

7. Jansson, P.E., Kenne, L. and Lindberg, B. (1975) Carbohyd. Res. 45,275-282.
8. Melton, L.D., Mindt, L., Rees, D.A. and Sanderson, G.R. (1976) Carbohyd. Res., 46,245-257.
9. Choy, Y.M. and Dutton, G.G.A. (1973) Can. J. Chem. 51,198-207.
10. Choy, Y.M., Fehmel, F., Frank, N. and Stirm, S. (1975) J. Virology 16,581-590.
11. Gorin, P.A.J., and Spencer, J.F.T. (1961) Can. J. Chem. 39, 2282-2289.
12. Gorin, P.A.J. and Spencer, J.F.T. (1963) Can. J. Chem. 41, 2357-2361.
13. Orentas, D.G., Sloneker, J.H. and Jeanes, A. (1963) Can. J. Microbiol., 9,427-430.
14. Lesley, S.M. and Hochster, R.M. (1959) Can. J. Physiol. 37, 513-529.
15. Arnott, S. and Scott, W.E. (1972) J. Chem. Soc. (Perkin II) 324-335.
16. Guss, J.M., Hukins, D.W.L., Smith, P.J.C., Winter, W.T., Arnott, S., Moorhouse, R. and Rees, D.A. (1975) J. Mol. Biol. 95,359-384.
17. Winter, W.T., Smith, P.J.C. and Arnott, S. (1975) J. Mol. Biol. 99,219-235.
18. Moorhouse, R., Winter, W.T. and Arnott, S. (1976) J. Mol. Biol. in press.
19. Smith, P.J.C. and Arnott, S. (1976) Acta Crystallogr., in press.
20. Arnott, S. (1973) Trans. Amer. Crystallogr. Assoc., 9,31-56.
21. Rees, D.A. (1973) In MTP International Review of Science: Organic Chemistry Series 1, vol. 7, G.O. Aspinall, ed. 251-283.
22. Arnott, S., Fulmer, A., Scott, W.E., Dea, I.C.M., Moorhouse, R. and Rees, D.A. (1974) J. Mol. Biol., 90,269-284.
23. Morris, E.R. and Rees, D.A. (1976) J. Biol. Chem., in press.
24. Holzworth, G. (1976) J. Biol. Chem., in press.
25. Dea, I.C.M., McKinnon, A.A. and Rees, D.A. (1972) J. Mol. Biol. 68,153-172.
26. Moorhouse, R. and Arnott, S., J. Mol. Biol., in preparation.
27. Morris, E.R. personal communication.

Infrared and Raman Spectroscopy of Polysaccharides

JOHN BLACKWELL

Department of Macromolecular Science, Case Western Reserve University,
Cleveland, OH 44106

During the last 30 years, infrared spectroscopy has been used to obtain information about the physical structures and chain conformations of polysaccharides. In recent years, the Raman spectra have also been available, and have provided useful complementary data. These techniques have mainly been applied in conjunction with other structural methods, especially x-ray diffraction, where the vibrational data have often given information on hydrogen bonding networks and side-group orientations.

This work for polysaccharides can be discussed in two general areas. Firstly, there are the direct structural investigations, which have utilized the identifiable group frequencies. The O-H, C-H, and carboxyl stretching frequencies, as well as some of the amide modes, can be identified and their infrared dichroisms determined. Hence, Marrinan and Mann [1] and subsequently Liang and Marchessault [2,3] showed that the four polymorphic forms of cellulose had different spectra in the O-H stretching region, indicative of different hydrogen bonding in their crystal structures. Based on the dichroisms of the O-H and C-H stretching bands, these authors discussed the possibilities for hydrogen bonding and selected what they considered the most likely structures. Similarly for chitin, [4,5] the orientation of the amide side chain relative to the fiber axis was determined from the dichroisms of the amide I and II bands. Secondly, known conformations of polysaccharides can often be differentiated by their I.R. and Raman spectra. Apart from the stretching frequencies listed above, most of the bands in polysaccharide spectra are due to complex molecular motions and structural interpretation of their dichroisms is not possible at this time. Nevertheless, despite this lack of understanding, changes in frequency or intensity can be used to follow polymorphic transitions. For example, the transition from cellulose I to cellulose II during mercerization has been followed by monitoring four intensities in the 1500-800cm^{-1} range [6,7].

This second aspect: identification of known conformations, is probably the major area for potential structural work on polysaccharides using this technique. Raman spectroscopy and the recently developed Fourier transform I.R. method, allow the spectra of polysaccharides in solution to be recorded at resolutions comparable to the solid state spectra. As a result, it is possible to compare the solution spectra with those of known solid state structures and hence assign a conformation to the polysaccharide in solution or in gels, in a manner analogous to identification of polypeptide conformations in solution using circular dichroism.

In this paper I will review some of the progress we have made in the last few years in analysis of a variety of polysaccharide systems. Our initial work on the polymorphic forms of amylose led on to studies of the spectra of oriented films of connective tissue glycosaminoglycans and hence to our present interest in bacterial polysaccharides in solution. In addition, we have made theoretical predictions of polysaccharide spectra using normal coordinate analysis.

Amylose

Amylose ($\alpha(1,4)$-$\underline{D}$-glucan) is the simplest polysaccharide which can be crystallized in different chain conformations. Precipitation from organic solvents leads to the so-called V-amylose structure, ($\underline{8,9}$) where the chains form compact helices with six glucose residues per turn repeating in 8.0Å. A variety of chain packings are possible, depending on the degree of hydration of the presence of organic solvent molecules, but the basic chain conformation is believed to be the same. When V-amylose is maintained at high humidity for a period of time, conversion occurs to one or other of the structures found in native starch, A- and B-amylose, which again are believed to be different packings of a common chain conformation. The proposed conformation for B-amylose ($\underline{10}$) is a more extended 6_1 helix, repeating in 10.4Å. Double helices have also been considered, ($\underline{11}$) but such structures will also involve more extended chains than occur in V-amylose.

The Raman spectrum of V-amylose ($\underline{12}$) is shown in Figure 1. The spectrum for B-amylose is very similar, except for four small but significant differences, which are shown in Figure 2: lines at 946 and 1263cm^{-1} for V-amylose shift to 936 and 1254cm^{-1} respectively in the B-form, and the relative intensities of lines at 1334 and 2940cm^{-1} are decreased with respect to their neighbors ($\underline{12}$). Based on our own C-H and O-H deuterium exchange experiments, three of the lines in question can be assigned as follows. The 2040cm^{-1} is probably a CH$_2$ antisymmetric stretching mode; those at 1334cm^{-1} and 1263cm^{-1} are mixed -CH$_2$OH deformation modes. For the mode at 946cm^{-1}, from a study of the spectra of glucose monomers and oligomers this is assigned as a linkage mode,

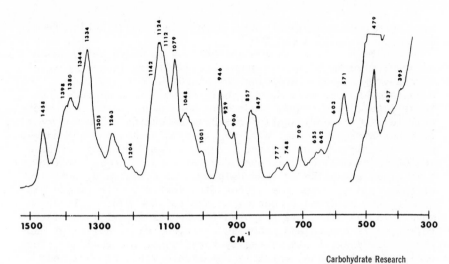

Carbohydrate Research

Figure 1.   Raman spectrum of V$_a$-amylose in the region 1500–300 cm^{-1} (12)

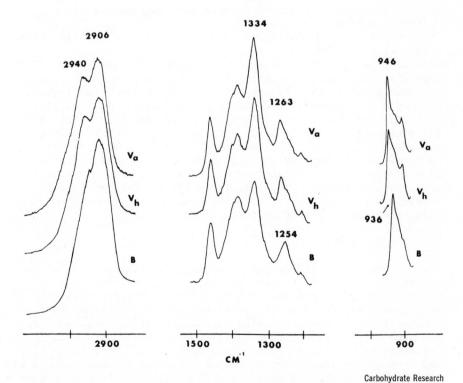

Carbohydrate Research

*Figure 2.   Comparison of regions of the Raman spectra of V$_a$-, V$_h$-, and B-amylose.
V$_a$- and V$_h$- are different hydrates of V-amylose; their spectra are identical (12).*

i.e. a complex mode involving a significant contribution from motion of the glycosidic C-O-C.

The V-structure has compact helices in which residues on successive turns (i.e. residues i and i+6) are linked by interturn hydrogen bonds involving the $-CH_2OH$ side chains. On conversion to the B-form, the chain becomes more extended and these interturn hydrogen bonds will be broken. This will probably lead to a reorientation of the side chains and the formation of other hydrogen bonds, e.g. to water molecules. Such changes would be likely to affect the frequency and intensity of the $-CH_2OH$ modes and would account for the changes seen. At the same time, expansion of the chain will be effected by rotation of the residues about the glycosidic linkages, which would fit in with the observed frequency change for the $946cm^{-1}$ linkage mode. In the structure proposed for B-amylose, (11) the interturn bond is broken and reformed through a water molecule, and the fiber repeat is increased by rotation of the residues about the glycosedic bonds, which is compatible with the observed Raman changes. Normal coordinate analysis of the isolated V-amylose chain predicts complex deformation modes which are in accord with the above assignments. (13) Increase in the fiber repeat of the helix to 10.4Å reduces the frequency of the "linkage" mode by 4 cm^{-1}.

The above Raman characteristics for V- and B- amylose can be used to interpret the spectra of this polymer in solution. Figure 3 shows the Raman spectrum of amylose in deuterated DMSO. (12) Only a short region of the spectrum can be recorded, but the spectral characteristics are those of the B-form, with the observed frequency at $1254cm^{-1}$ and relatively low relative intensity for the line at $1334cm^{-1}$. These results are against the presence of the V-helix in solution, which is interesting since the V-structure is formed when films are cast from this solvent. This is not to say that B-helices are present in solution since we believe that random, solvated amylose may show the same characteristics. However, it is likely that the CH_2OH groups are hydrogen bonded to solvent molecules rather than being involved in interturn bonds on compact V-helices.

Glycosaminoglycans

We are in the process of extending this type of work to the glycosaminoglycans of connective tissue, each of which can be prepared as oriented films in a number of different chain conformations, depending on the relative humidity and type of counter ions. In collaboration with E.D.T. Atkins and coworkers at University of Bristol, we have prepared crystalline film specimens of hyaluronic acid, chondroitin 4- and 6-sulfates, and dermatan sulfate. Raman spectra could not be obtained due to fluorescence of the specimens in the laser beam. However, using Fourier transform techniques we have been able to record the infrared spectra of the same oriented films (14) as were prepared for x-ray work. Figure 4 shows the polarized infrared spectra of chondroitin 6-sulfate, prepared in the 8_3 helical conformation. This polymer approximates to a repeating disaccharide of N-acetyl-D-galactosamine 6-sulfate and D-glucuronic acid, with alternating β(1,4) and β(1,3) linkages, and has eight disaccharides repeating in three turns, with a rise per residue of 9.8Å (15). The spectra in Figure 4 show perpendicular dichroism for the amide I and II modes at 1650 and 1560cm^{-1} respectively, indicating that the plane of the amide group is approximately perpendicular to the chain axis. Similarly the antisymmetric and symmetric carboxyl stretching frequencies at 1620 and 1420cm^{-1} respectively, both have slight perpendicular dichroism and the plane of the carboxyl group is more nearly perpendicular to the chain axis. The bands with parallel dichroism in the 1200-1000cm^{-1} range are complex C-O and C-C stretching modes. The dichroism is analogous to that for cellulose in the same range, and is characteristic of extended chain polysaccharides.

We have also prepared chondroitin 4-sulfate and dermatan sulfate, each in the 3_1 conformation, (16,17) and two forms of hyaluronic acid, both 4_1 conformations with different fiber repeats (18,19). These give similar results to those for chondroitin 6-sulfate for the amide orientation. For the two forms of hyaluronic acid, and chondroitin 4-sulfate however, the carboxyl symmetric stretching band has parallel dichroism. These conformations are less extended than the 8_3 form of C6S, and the C-COO bond can be oriented so that it is more nearly parallel to the chain axis. The same band has perpendicular dichroism for dermatan sulfate, which is consistant with the Cl chain for the L-iduronic acid residue of this polysaccharide (17).

So far we have only examined hyaluronic acid prepared in two different conformations, both 4_1 with different fiber repeats. These specimens do not show any spectral differences which can be ascribed to the difference in conformation. This is disappointing, but such differences are more likely when there are larger differences in conformation, e.g. between 3_1, 8_3, and 4_1 helices. These investigations are continuing, and will be applied to solutions if the different conformations can be successfully differentiated.

Bacterial Polysaccharides

More recently we have examined the bacterial polysaccharide xanthan, working with specimens obtained from Drs. A. Jeanes and P.A. Sandford at U.S.D.A., Peoria. This polysaccharide is believed to be a repeating polysaccharide, the backbone is a β(1,4)-glucan with alternating residues having a trisaccharide of mannose 6-acetate, glucuronic acid, and mannose; approximately 50% of the terminal mannose residues have a peruvate residue attached at the 4 and 6 positions.

Carbohydrate Research

Figure 3. Raman spectrum of amylose in deuterated DMSO solution ($Me_2SO_4 - d_6$) in the 1500–1200 cm^{-1} region. (– – –) indicates the approximate base line for scattering by the solvent (12).

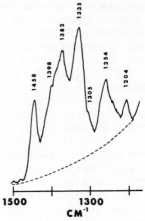

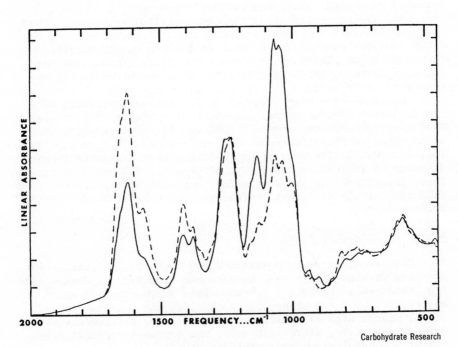

Figure 4. Polarized infrared spectra for the 8_3 conformation of chondroitin 6-sulfate.
(———) $A_{||}$; (– – –) $A_\perp$ (14).

Carbohydrate Research

This polysaccharide has a very interesting property in that the visiosity of an aqueous solution undergoes a sudden increase as the temperature rises (20). It is argued that the polysaccharide has a compact conformation at low temperatures and undergoes a transition to an expanded form at a specific temperature, reported at 55°C. This transition has been followed by Rees and coworkers (21) by N.M.R., which indicates an ordered conformation below 55°C and a disordered random coil at higher temperatures. These workers have also followed the change by circular dichroism spectroscopy.

We have used Fourier transform infrared spectroscopy to investigate these thermal changes (22). The specimens were dialyzed thoroughly against distilled water prior to recording the spectra. The spectra of a 1% xanthan solution at different temperatures are shown in Figure 5. No obvious differences are seen in the frequencies of the observed bands, but there is a general broadening at higher temperatures, indicating development of a less ordered state. This broadening can be quantized in a variety of ways; one convenient method is to measure the areas of the peaks above the unresolved background. Plots of these "intensities" against temperature for three of the bands are shown in Figure 6. All three show a sigmoidal transition, with midpoint at 40°C, indicating development of a more random conformation above this temperature.

Addition of salts to the xanthan solution is known to prevent the transition in the viscosity (20). Figure 7 shows the infrared spectra of a 1% xanthan solution in 1% KCl over the same temperature range as in Figure 5. The contrast between Figures 5 and 7 is quite striking in that the spectra of the salt solutions show very little change with temperature.

Our observations of a transition at 40°C is puzzling since other workers have reported 55°C. We have also performed viscosity and CD measurements on solutions of this polysaccharide, and observe transitions with midpoints of 38° and 40°. It is possible that our specimen of xanthan is different from those used by other workers, perhaps due to mutation or degradation, or that we have achieved a lower ionic strength when the specimen was dialysed against water.

Acknowlegements

This work was supported by N.S.F. Grant No. GB 32405. I am indebeted to my collaborators in Cleveland and Bristol, expecially J.J. Cael, J. Southwick, and J.L. Koenig for their part in the work described above.

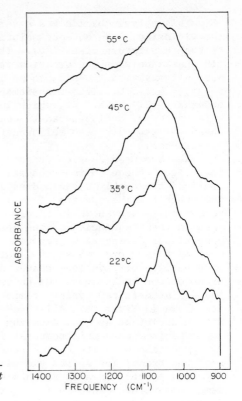

Figure 5.   Fourier transform infrared spectra of a 1% solution of xanthan in water at 22°, 35°, 45°, and 55°C (22)

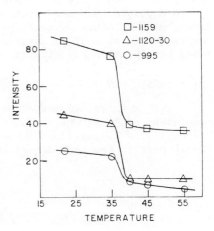

Figure 6.   Plots of the areas above the baseline for bands at 1159, 1120–1130, and 995 cm⁻¹ in the spectra Figure 5 (22)

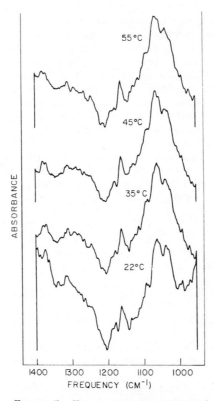

Figure 7. Fourier transform infrared spectra of a 1% solution of xanthan in 1% aqueous potassium chloride solution at 22°, 35°, 45°, and 55°C (22)

Abstract

Progress in several areas is described in the application of vibrational spectroscopy to investigate the structure and conformation of polysaccharides. Infrared and Raman spectroscopy provides information on the orientation of side groups and the type of hydrogen bonds formed in crystalline polysaccharide structures. In addition, spectral characteristics of polysaccharides prepared in different known crystal structures can be used to investigate the conformation in solution. These methods have been applied to investigations of amylose, where differences in the Raman spectra of the V- and B- forms have been interpreted in terms of the change in conformation, and indicate that the V-conformation is not present in solution. Fourier transform infrared spectra of oriented crystalline films of the connective tissue glycosaminoglycans have been used to determine the orientation of the amide and carboxyl groups for the various crystal structures. Finally, infrared spectra of xanthan in solution show that an order-disorder transition occurs as the temperature is increased, which is correlated with the sharp increase in viscosity in the same temperature range.

Literature Cited

1. Marinan, H.J. and Mann, J., J. Polymer Sci. (1958), 32, 357.
2. Liang, C.Y. and Marchessault, R.H., J. Polymer Sci., (1959), 37, 385.
3. Marchessault, R.H. and Liang, C.Y., J. Polymer Sci., (1960), 43, 31.
4. Darmon, S.E. and Rudall, K.M., Disc. Farad. Soc., (1950), 9, 215.
5. Carlstrom, D., J. Biophys. Biochem. Cytol., (1957), 3, 669.
6. McKenzie, A.W. and Higgins, H.G., Svensk Papperstidn., (1958) 61, 893.
7. Hurtubise, F.G. and Krassig, H., Anal. Chem., (1960), 32, 177.
8. Rundle, R.F. and French, D., J. Amer. Chem. Soc., (1943), 65, 558.
9. Zobel, H.F., French, A.D., and Hinkle, M.E., Biopolymers, (1967), 5, 837.
10. Blackwell, J., Sarko, A., and Marchessault, R.H., J. Molec. Biol., (1969), 42, 379.
11. Kainuma, K. and French, D., Biopolymers, (1972), 11, 2241.
12. Cael, J.J., Koenig, J.L., and Blackwell, J., Carbohydrate res., (1973), 29, 123.
13. Cael, J.J., Koenig, J.L., and Blackwell, J., Biopolymers, (1975), 14, 1885.
14. Cael, J.J., Isaac, D.H., Blackwell, J., Koenig, J.L., Atkins, E.D.T., and Sheehan, J.K., Carbohydrate Res. in press.
15. Arnott, S, Guss, J.M., Hukins, D.M., and Mathews, M.B., Science, (1975), 180, 743.

16. Isaac, D.H. and Atkins, E.D.T., Nature (London), New Biol. (1973), 244, 252.
17. Atkins, E.D.T. and Isaac, D.H., J. Molec. Biol., (1973), 80, 773.
18. Dea, I.C.M., Moorhouse, R., Rees, D.A., Guss, J.M., and Balazs, E.A., Science, (1973), 179, 560.
19. Guss, J.M., Hukins, D.W., Smith, P.J.C., Winter, W.T., Arnott, S., Moorhouse, R., and Rees, D.A., J. Molec. Biol., (1975), 95, 359.
20. Jeanes, A., Pittsley, J.E., and Senti, A., J. Appl. Polymer Sci., (1961), 17, 519.
21. Rees, D.A. and Morris, E., (in press).
22. Southwick, J., Koenig, J.L., and Blackwell, J., (in press).

9

Nuclear Magnetic Resonance and Mass Spectroscopy of Polysaccharides

FRED R. SEYMOUR

Baylor College of Medicine, Marrs McLean Department of Biochemistry,
Houston, TX 77030

Various physical methods for polysaccharide structural determination have been developed. This discussion will be primarily limited to gas-liquid chromatography/mass spectrometry (g.l.c.-m.s.) and nuclear magnetic resonance spectrometry (n.m.r.). The equipment employed for these determinations has either been recently developed, or recently brought to the degree of sophistication necessary for carbohydrate studies. The procedures are rapid compared to previous methods of obtaining analogous data. Though most of the techniques may be applied to any carbohydrate containing compound, this work has been done with extra-cellular polysaccharides. The extra-cellular polysaccharides have proven to be valuable materials as they are readily obtainable homogeneous polymers which can be produced in relatively large quantities. These large amounts of uniform polysaccharides provide material to compare the g.l.c.-m.s. data to the much less sensitive (in terms of amount required) n.m.r. data. The wide variety of mannans and glucans available provide a considerable range of structures for this correlation. Though data is available from a number of different sources, the selected examples will be chosen from studies in which I have participated.

Five general methods have been employed: a) polymer hydrolysis followed by g.l.c.-m.s., b) polymer permethylation followed by hydrolysis and then g.l.c.-m.s., c) recording the polymer's C-13 n.m.r. spectra, d) recording the polymer's P-31 n.m.r. spectra, and e) employing selective hydrolysis combined with high pressure chromatography (h.p.c.) These methods are complimentary in duplication and confirmation of data, but each method yields specific unduplicatable information.

Gas-liquid chromatography

A variety of derivatives, column packings, and oven

114

conditions have been employed for saccharide g.l.c. separation(1). In our hands, the peracetylated aldononitriles (PAAN) have proven to be the most useful derivatives. The basic reaction for conversion of saccharides to their PAAN derivatives is well known. Differing g.l.c. conditions have been reported for the separation of PAAN derivatives (2, 3, 4). We find a most satisfactory system is to use neopentyl glycol succinate. The resulting narrow peaks allow good compound separation, the columns can easily handle the amounts of material required for m.s. determinations, and the base line remains flat, not affecting the hydrogen flame detector or the mass spectrometer.

The advantages of the PAAN derivatization procedure are, a) it is fast and efficient, b) the anomeric center of asymmetry is destroyed, each saccharide yielding a single derivative, and c) the resulting straight-chain derivatives have readily interpretable mass spectra.

It is assumed that a successful non-degrading hydrolysis has preceded the saccharide derivatization and g.l.c.-m.s. determination. Carbohydrate hydrolysis is essentially a subject in itself, but it can be noted that the C-13 n.m.r. data provide a useful non-destructive check on g.l.c.-m.s. structural determinations which hypothesize a non-degrading hydrolysis.

TABLE 1

Retention Times of Peracetylated Aldononitrile Derivatives of Aldoses (a)

Parent Aldose	Retention Time (min)	Parent Aldose	Retention Time (min)
DL-glyceraldehyde	1.2	D-allose	19.0
D-erythrose	5.6	2-deoxy-D-glucose	19.2
D-digitoxose	9.6	D-mannose	19.6
L-rhamnose	10.6	D-talose	19.6
2-deoxy-D-ribose	12.2	2-deoxy-D-galactose	20.2
D-ribose	12.8	D-glucose	21.0
L-fucose	12.8	D-galactose	21.8
D-Lyxose	13.6	5-thio-D-glucose	24.6
D-arabinose	14.2	D-glucoheptose	26.8
D-xylose	15.6	N-acetyl-D-glucosamine	34.2

(a) 3% Neopentyl glycol succinate on 60/80 mesh Chromosorb W in a packed glass column 2 mm by 4 ft.

The retention times of 20 saccharide PAAN derivatives are shown in Table 1. Only two pairs cannot be resolved and only one differs only in terms of stereochemistry (mannose and talose) and cannot be separated by g.l.c.-m.s. This allows one reasonable confidence in establishing which sugars are not present. Partial

lists of the retention times of these PAAN derivatives on different columns (packed LAC-4R-886 (3) and open tubular SE-30 (4))
indicate that the retention time order is the same for these
compounds. The retention times of these PAAN derivatives appear
to be a function of the interaction between the acetyl groups
and the stationary phase. The degree of acetyl-stationary phase
interaction is apparently dependent on the number of acetyl
groups per molecule, and the availability of these groups to
interact with the stationary phase. For the PAAN derivatives of
the unsubstituted saccharides (triose, tetrose, pentoses, etc.)
the order of emergence occurs in groups: first triose (glyceraldehyde), then tetroses (e.g. erythrose), pentoses (e.g. ribose),
hexoses (e.g. mannose), and heptose (glucoheptose) PAAN derivatives. In addition, for tetroses the erythrose PAAN derivative's
retention time has been found to be smaller than the threose PAAN
derivative, for pentoses the ribose PAAN derivative's retention
time is smallest and the xylose PAAN derivative's retention time
is the largest (4 , 5)`. For each class of stereoisomers (e.g.
the pentoses) the stereoisomer containing the greatest number of
pairs of cis acetyl groups (or hydroxyl groups in the underiva-
tized sugar) has the smallest retention time, and the stereoisomer
containing the smallest number of pairs of cis acetyl groups the
largest retention time. It is possible that the cis acetyl groups
promote saccharide chain bending, and these less linear molecules
provide less opportunity for acetyl groups to interact with the
stationary phase. On the basis of limited data, it appears that
the replacement of a functional groups uniformly changes the
retention time of a series of stereoisomers. For example, the
2-deoxy-D-glucose and 2-deoxy-D-galactose PAAN derivatives have
retention times of 1.7 minutes less than their respective D-
glucose and D-galactose PAAN derivatives and the 6-deoxy-L-mannose
and 6-deoxy-L-galactose PAAN derivatives have retention times
9.0 minutes less than their corresponding mannose and galactose
PAAN derivatives. It is possible that on 6-deoxy hexose
substitution the remaining 4 acetyl groups have the same general
acetyl-stationary phase interaction as the pentoses with the 6-
methoxy group not participating. However, in the case of 2-
methoxy hexose substitution (or any non-terminal substitution) the
methylene unit acts as a chain extender with the acetyl-stationary
phase interaction still approximating that of a normal hexose
PAAN derivative.

      Under the standard PAAN derivatization procedure the 5-thio-
D-glucose results in a well defined g.l.c. peak. However, m.s.
shows that this g.l.c. peak is not the PAAN derivative, but the
peracetylated 5-thio-D-glucopyranoside. N-acetyl-D-glucosamine
yields the corresponding PAAN derivative with a retention time
much longer than the D-glucose PAAN derivative -- indicating
increased N-acetyl interaction with the stationary phase.

Mass Spectrometry

The ammonia chemical ionization (c.i.a.) - m.s. of the above PAAN derivatives are very simple. The most prominent, and usually only, m/e peaks are M + 18 and M - 59, representing addition of the ammonium ion, or the addition of a proton and successive loss of acetic acid. The derivatization procedure is normally duplicated, first with N-15 hydroxylamine and then with per-deuterated acetic anhydride. The N-15 introduction shifts M by 1 a.m.u. for each aldehyde originally present and the deuterium shifts M by 3 a.m.u. for each hydroxyl group originally present. Therefore, the molecular weight, number of hydroxyl groups, and number of aldehyde groups present in aldoses can rapidly be established.

The electron impact (e.i.) - m.s. yields results from carbon-carbon cleavage of the carbohydrate backbone chain. This backbone cleavage is equally likely between any carbons, except the C-1 and C-2 positions, and different length fragments are generated from both ends. The glyceraldehyde PAAN derivative gives very few m/e fragments; these same m/e also appear in the erythrose PAAN spectra with a new set of m/e fragments. As the molecule is lengthened, more fragments are possible, and comparison of the spectra of different length molecules indicates the original carbohydrate position of each fragment. The fragmentation pathways are also identified by N-15 nitrile substitution and by deuteroacetyl substitution.

Upon the substitution of a functional group at a specific position in the carbohydrate molecule, all m/e fragments originating from that position will be shifted. Therefore, the functional groups mass and position can be established.

For an aldose PAAN derivative, a combination of g.l.c.-m.s. using c.i.a. and e.i. can establish the molecular weight, the number of aldehyde and hydroxyl groups, the type and position of functional groups, and provide an estimate of the stereochemistry of the molecule.

Methane chemical ionization (c.i.m.) - m.s. of PAAN derivatives have been examined and though interpretable spectra are obtained for each compound, if the c.i.a.-m.s. and the e.i.-m.s. are known, little additional information is obtained. In general, c.i.m. mass fragments result from the progressive and extensive loss of functional groups (the O-acetyl groups are lost as acetic acid and ketene) and the backbone chain remains intact. For the PAAN derivative of N-acetyl glucosamine, the N-acetyl group is not easily lost and therefore the spectrum of this compound is very similar to the corresponding glucose PAAN derivative -- with each m/e fragment decreased by 1 a.m.u.

Permethylation Gas-Liquid Chromatography/Mass Spectrometry

Polysaccharide permethylation, followed by hydrolysis and

analysis of the resulting methyl ether saccharides has traditionally been employed for determining sugar-sugar linkage type and
degree of branching. The procedure is simple in concept, but has
proven difficult to apply. This is due to the difficulty of
identifying and quantitating the reaction products, or their
derivatives, and has been used in conjunction with peracetylated
cyclic derivatives or the peracetylated alditols. Lance and
Jones (6) separated methyl ethers of xylose PAAN compounds and
Dmitriev et al. (2) reported the major m/e fragments of the e.i.-
m.s. of selected PAAN derivatives. The tetra-0, tri-0, and
di-0-methyl ethers of D-mannopyranoside were synthesized
and g.l.c. conditions found for the separation of these
compounds (7). This g.l.c. separation, combined with previously
developed efficient methylation and hydrolysis methods, allowed
the rapid permethylation analysis of a series of mannans (8). The
identity of the g.l.c. peaks can be easily determined by e.i.-
m.s. The availability of the pure methyl ethers of methyl α-D-
mannopyranoside and the establishment of a g.l.c. column
capable of resolving the PAAN derivatives, allowed a precise
determination of the fragmentation pathways. Di-0-methyl
derivatives of methyl α-D-mannopyranoside were subjected to
random partial methylation and the resulting reaction mixture
hydrolyzed and derivatized to PAAN derivatives. On g.l.c.
analysis these mixtures gave a series of peaks, the retention
times indicating the position of the combined methyl-0- and
deuteromethyl-0- ether groups. The position of new ether groups
identified these as deuteromethyl groups, and on comparison of
the e.i.-m.s. of these compounds to the e.i.-m.s. of the
corresponding non-isotopically substituted PAAN derivatives, the
origen of each m/e fragment could be established (7).
    The extensive knowledge of g.l.c.-m.s. allowed the structure
of sixteen mannans to be established in terms of linkage type
and degree of branching. This method yielded essentially
identical data for several of the mannans. Six classes of
mannans were observed, these polysaccharides differing in both
linkage types and degree of branching. This data is summarized
in Table 2.
    The data in Table 2 allows the construction of an average
repeating unit for each polymer class. For example, D-mannan
produced by Pachysolen tannophilus Y-2460 can be expressed as:

$$-\{M - (1{\rightarrow}6)\}-_x$$

$$\overset{|}{\underset{3}{}}$$
$$\uparrow$$
$$\underset{1}{}$$
$$|$$

$$M - (1{\rightarrow}2) - M - (1{\rightarrow}2) - M$$

TABLE 2

Mole Percentage of Methylated D-mannose Components in
Hydrosylates of Permethylated D-mannans (a)

NRRL Number	Methyl ethers of D-mannose					
	2,3,4,6-	3,4,6-	2,4,6-	2,3,4-	3,4-	2,4-
Y-1842	2.5	51.1		44.3	2.0	
Y-2448	3.9	48.3	44.0		3.9	
Y-2460	26.4	42.9		2.1		26.1
Y-2023	19.2	54.1		9.7	16.9	
YB-2097	27.7	20.4	25.2			23.8
YB-1344	22.8	4.2	48.0	2.6	2.3	20.1

(a)   Data taken from reference 8.

As a first approximation, it was assumed that the (1→6)-
linkages were exclusively confined to the mannan backbone chain.
This assumption was later tested by acetolysis (see below).
This g.l.c.-m.s. technique was then applied to glucans.
The generation of reference compounds was not necessary. The
mass spectra are not affected by stereochemistry changes, and
the corresponding mannose and glucose methyl-ethers yield
identical spectra. Therefore, each glucose methyl ether PAAN
derivative g.l.c. peak could be identified by comparison to the
known mannose compounds. On the butanediol succinate columns
employed for methyl-ether saccharide PAAN separation the
retention times are generally, but not necessarily, different
for corresponding glucose and mannose compounds. A group of
dextrans, previously suspected of containing unusual structural
features, was analyzed by g.l.c.-m.s. and the results (9) are
summarized in Table 3.
The data in Table 3 again allows the construction of
average repeating units for the various polysaccharides. For
example, fraction L of the dextran produced by Leuconostoc
mesenterodies NRRL B-1299 can be expressed as having a general
repeating unit of:

$$-\{G - (1{\to}6) - G - (1{\to}6)\}-_x$$
$$|$$
$$\widehat{2}$$
$$\uparrow$$
$$\underset{|}{\underline{1}}$$
$$G$$

where "G" is the D-glucopyranoside unit.

TABLE 3

Mole percentages of Methylated D-glucose Components in
Hydrolyzates of Permethylated Dextrans (a)

NRRL Number (b)	Methyl ethers of D-glucose						
	2,3,4,6-	2,3,4-	2,3,6-	2,4,6-	2,3-	2,4-	3,4-
B-1351 S	5.8	83.3				10.5	0.3
B-1399 L	12.8	74.5				5.9	6.8
B-1254 L	22.1	55.0	3.4		19.5		
B-1299 S	39.1	26.0					34.9
B-1355 S	6.9	46.9		35.0		11.2	

(a)   Data taken from reference 9.
(b)   The dextran producing NRRL strain number. S and L refer to
      polymer fractions.

     At this point it will be seen that g.l.c.-m.s. has been
used to confirm the unique presence of glucose or mannose as
the aldose unit of a series of dextrans and a series of mannans.
In conjunction with permethlylation, g.l.c.-m.s. has been
employed for establishing the general repeating unit for these
dextrans and D-mannans. A number of D-mannans, not listed in
Table 2, were shown to have essentially identical linkage
types to those shown. It is possible, in principle, to perform
these operations on a sub-miligram basis. For ease of materials
handling, a few mg of each polymer were employed for the mannan
determinations, and due to increased permethylation difficulty,
approximately 10 to 15 mg of dextrans were used. This data then
provided the basis for comparison with the remaining techniques,
which require larger amounts of material.

High-Pressure Chromatography

     Previous work has shown that acetolysis of mannans results
in selective hydrolysis, with the (1→6)-linkages cleaved much
more rapidly than other sugar-sugar linkages (10). Employing
this selective hydrolysis, followed by a deacetylation step,
yielded a mixture of oligosaccharides. It had previously
proven possible to employ h.p.c. to separate a mixture of
oligosaccharides.   By calibrating the system against known
oligosaccharides, the retention times and detector responses
could be established. The h.p.c. system was then employed to
separate and quantitate the acetolysis oligosaccharides according
to degree of polymerization (d.p.).   An example of this data for
the mannan of Pachysolen tannophilus, NRRL Y-2460 is shown in
the following table.

TABLE 4

Oligosaccharides from Acetolysis of NRRL Y-2460 D-mannan

Degree of Polymerization	1	2	3	4	5	6	7	8
Mole ratio	10.0	3.7	3.6	11.9	1.3	0.7	1.0	0.9

This data is analyized by making two basic assumptions, a) only the (1→6)-linkages have been broken, and b) all the (1→6)-linkages are in the carbohydrate backbone. If this is correct, then sequences of (1→6)-linkages will yield monomers, and the side chains will remain unaffected and attached to a single backbone saccharide unit. Therefore, each oligimer will represent a side chain non-reducing end group, a branching end group, and non-(1→6)-linked saccharides. This data may then be analyzed to yield an average repeating unit which can be expressed in terms of methyl ethers -- an example is shown below.

TABLE 5

Correlation of Methylation and Acetolysis Data for NRRL Y-2460 D-mannan (a)

Data Source	Calculated percentages of methyl ethers			
	Tetra	2,3,4-Tri	Non-2,3,4-Tri	Di
Methylation	26.4	2.1	42.9	26.1
Acetolysis	23.0	10.1	44.5	23.0

(a) Data taken from reference 8.

It can be seen that the acetolysis data closely parallels the methylation data. The major discrepancy is the amount of (1→6)-linkages (2,3,4-tri-0-methyl ether) as acetolysis generally gives a higher value than methylation. The (1→6)-linkage acetolysis value comes from the amount of monomeric units observed by h.p.c., and it is possible that non-(1→6)-linkage cleavage contributes to increase this value. Two general results are obtained from comparison of this data: firstly, the assumption that the (1→6)-linkages are confined to the backbone is confirmed; secondly, .there appears to be a distribution of side chain lengths around the average of three saccharide units per side chain. Good agreement between acetolysis and methylation data was obtained for four of the six mannan types studied. Mannan Y-1842 showed great differ-

ences in degree of branching as determined by methylation and
acetolysis. The data suggest that a large number of (1→6)-
linkages must occur in the side chains.

For polysaccharides containing (1→6)-linked saccharides, the
correlation of methylation and acetolysis data is a useful method
to establish the position of the (1→6)-linkages. If these (1→6)-
linkages form the backbone chain, the side chain length
distribution can be established.

C-13 n.m.r. Spectroscopy

This technique is a logical step after the constituent
sugars, linkage types, and degree of branching have been established by g.l.c.-m.s. In principle, each carbon in a different
chemical environment will display a unique chemical shift in the
C-13 spectral region. The separation of the C-13 n.m.r. chemical
shifts is dependent on the magnetic field strength. When compared
to H-1 n.m.r., C-13 n.m.r. gives much better separations in an
equivalent field. In addition, due to improved relaxation times,
C-13 n.m.r. can give quite sharp signals for large polymers. It
has been established that simple saccharides (e.g. methyl α-D-
glucose) will yield six C-13 n.m.r. saccharide peaks; the
anomeric carbon in the 95 to 105 ppm (relative to TMS) region,
the C-2, C-3, C-4, and C-5 peaks in the 70-75 ppm region, and the
C-6 peak at approximately 60 ppm (11). On conversion of a
hydroxyl group to an alkyl ether group, the chemical shift of the
corresponding saccharide carbon is shifted downfield (to larger
ppm values). For methyl ether formation this change in chemical
shift has been shown to be a uniform 10 ppm downfield shift for
each saccharide carbon position (12). Therefore, a convenient
approach to polysaccharide analysis is to consider the polymer as
an agregation of independent alkyl ether monosaccharides. For
example, methylation data (Table 3) and the implied general
repeating unit have been presented for dextran B-1299 fraction S.
For purposes of C-13 n.m.r. this proposed basic unit may be
considered as equivalent to an equal molar mixture of methyl D-
mannopyranoside (the end group), methyl 2,6-di-O-methyl-D-manno-
pyranoside (the branching group), and methyl 6-O-methyl-D-manno-
pyranoside (the backbone extending group). For three saccharides,
a maximum of eighteen (6x3) saccharide C-13 chemical shifts could
be observed.

Two limitations of C-13 n.m.r. should be recognized. First,
the relatively low sensitivity of the C-13 nuclei requires large
samples (100 to 200 mg) with Fourier transform data processing.
Smaller samples may be used, but the data acquisition time
steadily increases. Secondly, the signal intensity of each class
of carbon nuclei is not dependent on the total number of each
species present. C-13 nuclei with greater degrees of freedom
of motion yield larger signals. However, it has previously been
shown that for saccharides, the contribution of each carbon

species to the C-13 n.m.r. spectra is <u>approximately</u> equal (<u>13</u>).
    For the C-13 n.m.r. spectra of dextran B-1299 fraction S
(Fig. 1), and analagous dextran spectra, several points become
apparent. For the dextrans described in Table 3, a series of
six chemical shifts (designated as A through F) are present in
each spectrum. These six peaks dominate the spectra of the more
linear α-linked dextrans and represent the contribution of a
methyl 6-0-methyl-D-glucopyranoside analog. As polymers with
greater degree of branching were examined, the contribution of
the original six peaks decreased and other chemical shifts become
prominent (<u>14</u>). B-1299 fraction S dextran (Fig. 1) provides
an example of a highly branched dextran with the contribution of
the original six peaks indicated by letters A through F. The
70-75 ppm region (B through D) could display twelve (3x4) chemical
shifts, but only seven are observed by electronic slope change
detection. Apparently a number of these chemical shifts are not
resolved. In the anomeric region the expected three peaks are
observed, the downfield peak representing the (1→6)-linked unit.
All anomeric protons are located in the 96 to 101 ppm region,
demonstrating that <u>each</u> of the observed linkages is α.
    For linear dextran the 75-85 ppm region displays no chemical
shifts. For dextrans containing 1→2, 1→3, or 1→4-linkages (as
demonstrated by g.l.c.-m.s.) the 75-85 ppm region contains the
glycosyl linked carbons (C-2, C-3, or C-4) which upon substitution have had their chemical shifts moved downfield from the 70-75
ppm region. We have observed a chemical shift of 76.5 ppm for
α-(1→2)-linkages, 79.5 ppm for α-(1→3)-linkages, and 81.6 ppm for
α-(1→4)-linkages (<u>14</u>). The only chemical shift in the 75-85 ppm

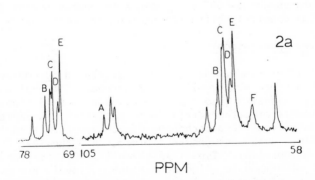

*Figure 1. C-13 NMR spectra of dextran B-1299 fraction S
recorded at 27° in D₂O; ppm relative to TMS. Inset (78–69
ppm) recorded at 70°.*

region of the dextran B-1299 fraction S spectra is at 76.42 ppm, in agreement with the g.l.c.-m.s. data. This n.m.r. method therefore allows identification and rough quantitation of the linkage types present by a non-destructive technique not dependent on hydrolysis.

In general, the dextran C-13 n.m.r. spectra were relatively simple, this being especially noticable in the anomeric region. A previous C-13 n.m.r. study of pulullans observed three well-defined anomeric chemical shifts and employed this as a very plausible argument for the ordered repeating $-(1\rightarrow4)-(1\rightarrow6)-(1\rightarrow4)-$ glucopyranoside subunit (15). Many of these dextrans also show simple C-13 n.m.r. spectra which in turn implies a basic ordered repeating sub-unit.

Another point of interest is the dependence of carbohydrate C-13 n.m.r. spectra on temperature. We noticed that branched polysaccharides had broad peaks which could be "sharpened" by raising the temperature. A high temperature (70°) inset of the 70-75 ppm region is shown in Figure 1. The general spectrum profile remains the same, but each peak is narrower. A number of temperature dependent effects were observed, the most interesting being that all of the chemical shifts were temperature dependent, moving downfield on increasing temperature. In addition, different chemical shifts displayed different temperature dependencies, in the range of $\Delta\delta/\Delta T$ of 0.01 to 0.03 ppm/C° (relative to TMS). These different $\Delta\delta/\Delta T$ exclude bulk magnetic susceptibility as the major factor and suggest that the magnitude of $\Delta\delta/\Delta T$ is structure related. In fact, the largest $\Delta\delta/\Delta T$ observed are generally associated with carbons involved in sugar-sugar linkages (14).

It is necessary to consider the $\Delta\delta/\Delta T$ effect, especially when comparing C-13 n.m.r. carbohydrate spectra in the closely packed 70-75 ppm region. The chemical shifts of the C-2, C-3, C-4, and C-5 carbons falling in this region are apparently diagnostic for saccharides of specific linkage types; however, a temperature change of 50° can cause a resonance change so great as to allow chemical shifts to interchange positions.

P-31 n.m.r. Spectroscopy

Additional n.m.r. data has been generated by employing a Fourier transform n.m.r. with a P-31 probe. The P-31 nuclei are relatively insensitive to n.m.r. and, as with C-13 studies, larger amounts of carbohydrates were necessary. A variety of extracellular yeast 0-phosphonohexosans were available for study. These compounds can be divided, on a chemical basis, into two groups (16). Type I is exemplified by poly(phosphoric diesters) of D-mannose oligosaccharides. Type II are polysaccharides in which the glycosyl phosphate residues occur as non-reducing end-groups -- either as D-mannose, D-glucose, or as disaccharides. Many of the mannans and phosphonohexoglucans

are related insofar as they are produced by the same yeast strain
under differing amounts of orthophosphate in the culture media.
        The P-31 n.m.r. signals from these 0-phosphonohexosans were
quite sharp. The native polymers apparently contain all phosphate
groups as the diester, but isolation procedures can result in
partial hydrolysis to the mono-ester. P-31 n.m.r. provides an
excellent method of surveying for this hydrolysis as the n.m.r.
signals of the mono-ester and the di-ester phosphates are widely
spaced, the mono-ester falling at approximately -4. ppm (relative
to 85% orthophosphoric acid).
        In general, each 0-phosphonohexoglycan gave a single sharp
P-31 signal, the chemical shift being unique for each polymer.
The two types of 0-phosphohexoglycans can be represented as:

$$- (M - M - M - P - M)_x -$$

$$- (\underset{\underset{\underset{G}{|}}{\overset{|}{P}}}{M} - M - M - M)_x -$$

Type I   Type II

Where M represents a mannopyranoside unit, P represents a
phosphodiester unit, and G represents a non-reducing mannose,
glucose, or disaccharide unit.
        Though each 0-phosphonomannan studied has displayed a
different P-31 chemical shift, there is no obvious difference
between Type I and Type II P-31 n.m.r. spectra (see Table 6).

TABLE 6

P-31 Chemical Shifts for 0-phosphonomannans and Related
Materials (a)

Anomeric sugar phosphate	NRRL producing strain	Orthophosphate diester chemical shift
	Type I	
mannose	Y-1842	1.94
"	YB-1443	1.74
"	Y-2448	1.84
"	Y-2461	1.72
	Type II	
mannose	Y-411	1.78
"	YB-2079	1.74 and 1.90
glucose	YB-2194	1.07
"	Y-2579	1.16
"	Y-2023	1.10 and 1.28
galactose	Y-6493	1.06

(a) data in Table 6 taken from reference 17.

However, Table 6 does show that a significant change in the chemical shift occurs when the anomeric sugar phosphate is glucose (at approx. 1.1 ppm) rather than mannose (at approx 1.8 ppm). The anomeric sugar phosphate is believed to be α-linked in all cases and the changes in chemical shift are the result of stereoisomer effects at the distance of several atoms from the phosphorus atom.

Relatively subtle changes in polymer structure are reflected in the P-31 chemical shift values. Polymers Y-2448 ($\delta = 1.84$ ppm) and Y-1842 ($\delta = 1.74$ ppm) are apparently structurally identical except that for Y-1842 the sugars are α-linked (not the sugar phosphate anomeric linkage) and for Y-2448 the sugars are β-linked. O-Phosphonomannan diesters containing mono- or di-saccharide residues also show different chemical shifts. For example, in polymers from YB-2097 and Y-2023, where both types of residue are present, two diester resonances are observed. In YB-2097 O-phosphonomannan the mannose 6-phosphate residues are in anomeric linkage with residues of mannopyranose and 6-O-α-D mannopyranosyl-D-mannopyranose, in Y-2023 O-phosphomannan the linkage is to residues of D-glucopyranose and 2-O-α-D-mannopyranosyl-D-glucopyranose.

Proton-coupled spectra for the diesters showed quartet patterns that could be analyzed by computer simulation to obtain the coupling-constants. These data are in accord with the interpretation that most of the linkages in the O-phosphono-mannans are of the D-mannopyranose 6-(D-mannopyranosylphosphate) type.

Conclusions

Examples have been presented to demonstrate how various forms of g.l.c., h.p.c., m.s., and n.m.r. are employed in extra-cellular polysaccharide structure determination. These structural determinations have proved fruitful in providing insights into the relationship of a wide variety of extra-cellular polysaccharides. In turn, the extracellular poly-saccharides have provided materials to correlate the various structural determination techniques, and these methods may now be applied to more complex saccharide containing polymers.

Literature cited

1. Dutton, G.G.S., Advan. Carbohyd. Chem. Biochem., (1974) 30, 9-110.
2. Dmitriev, B.A., Backinowsky, L.V., Chizhov, O.S., Zolotarev, B.M., and Kochetkov, N.K., Carbohyd. Res., (1971) 19 432-435.

3. Varma, R. Varma, R.S., and Wardi, A.H., J. Chromatogr. (1973) 77, 222-227.
4. Szafranek, J., Pfaffenberger, C.D., and Horning, E.C., Anal. Lett. (1973) 6, 479-492.
5. Seymour, F.R., Chen, E.C.M., and Bishop, S.H., Carbohyd. Res. in press.
6. Lance, E.G., and Jones, J.K.N., Can. J. Chem. (1967) 45, 1995-1998.
7. Seymour, F.R., Plattner, R.D., and Slodki, M.E., Carbohyd. Res. (1975) 44, 181-198.
8. Seymour, F.R., Slodki, M.E., Plattner, R.D., and Stodola, R. M., Carbohyd. Res. (1976) 48, 225-237.
9. Seymour, F.R., Slodki, M.E., Plattner, R.D., and Jeanes, A., Carbohyd. Res. in press.
10. Rosenfeld, L., and Ballou, C.E., Carbohyd. Res. (1974) 32, 287-298.
11. Perlin, A.S., Casu, B., and Koch, H.J., Can. J. Chem. (1970) 48, 2596-2606.
12. Usui, T., Yamoka, N., Matsuda, K., Tuzimura, K., Sugiyama, H. and Seto, S., J. Chem. Soc. Perkin, I, 1973, 2425-2432.
13. Gorin, P.A.J., Can. J. Chem., (1973) 51, 2375-2383.
14. Seymour, F.R., Knapp, R.D., Bishop, S.H., Carbohyd. Res. in press.
15. Jennings, H.J., and Smith, I.C.P., J. Am. Chem. Soc. (1973) 95, 606-608.
16. Slodki, M.E., Ward, R.M., Boundy, J.A., and Cadmus, M.C. in Terui, G. (Ed.), Proc. Int. Ferment. Symp. IVth: Ferment. Technol. Today, Soc. Ferment. Technol., Osaka, 1972 pp 597-601.
17. Costello, A.J.R., Glonek, T., Slodki, M.E., Seymour, F.R., Carbohyd. Res. (1975) 42, 23-37.

Acknowledgements

This work was supported, in part, by a Robert A. Welch Foundation Grant (Q 294), a National Science Foundation Grant (BMS-74-10433), and National Institutes of Health Grants (HL-05435, HL-14194, HL-17372). Special thanks are due to Drs. Allene Jeanes and Morey E. Slodki of the Northern Regional Research Laboratory, ARS, USDA, Peoria, Illinois, for providing the dextrans, mannans, and O-phosphonohexosans described in this paper.

10

Polysaccharide Polyelectrolytes

W. M. PASIKA

Chemistry Department, Laurentian University, Sudbury, Ontario, Canada

Macromolecules which possess a large number of some functionality and ionize in aqueous media are called polyelectrolytes. Ionization of the attached function aids in the solubilization of the polyelectrolyte substance and is responsible for its unique properties. Although the ionogenic function may be regarded as a salt, dissolution of the polyelectrolyte substance is not comparable to the dissolution of a simple salt. A simple salt such as sodium chloride in solution produces a cation and an anion of comparable size. Each ion has independent mobility. A polyelectrolyte dissolves to yield a polyion and counter ions. The polyion holds a large number of charges in close proximity because they are attached to the macromolecular backbone. Although the polyion has mobility, the individual charges attached to the chain do not. They remain within the domain of the macromolecular coil. Not all the gegions or counterions are completely mobile. Anionic polyelectrolytes have positive counter ions whereas cationic polyelectrolytes have negative counter ions. Polyampholytes can acquire either positive or negative charge along the macromolecular backbone depending upon the composition of the solution. Pictorially, one has the following

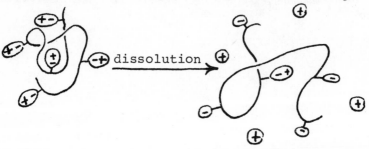

Because of free energy restrictions, not all the
ionogenic groups "ionize".   Many exist as ion pairs.
    A large number of polysaccharide polyelectrolytes
can be isolated from a variety of natural sources.
heparin, hyaluronic acid, chondroitin and keratin, to
name a few, are isolated from animal sources.   The
more familiar examples supplied by the plant world
are pectinic acids, alginates and carageenan.   A
number of polysaccharide polyelectrolytes, such as
Xanthan, can be obtained from nonpathogenic micro-
organisms (1).   The common characteristic is that the
macromolecular backbone is composed of saccharide
residues carrying ionogenic groups.   The latter are
more often than not carboxyl or sulfate functions.
"Synthetic" polysaccharide polyelectrolytes can be
obtained by suitably derivatizing polysaccharides.
The ensuing discussion will focus on derivatized
dextran in an attempt to illustrate some of the
factors which influence the characteristics of polysaccharide polyelectrolytes.

Viscosity.

    All macromolecular substances in solution
enhance the viscosity of the solvent considerably.
The larger the molecular weight or macromolecular
size, the greater the enhancement.   In characterizing
the macromolecular size through the viscosity enhancement, it is more conveniently done with the viscosity
functions listed in Fig. 1.   The dependence of
reduced viscosity on concentration of neutral macromolecular substances (i.e., dextran) is linear as
depicted in Fig. 1.   Extrapolation of the viscosity
data to "zero" concentration yields the intrinsic
viscosity, which measures the hydrodynamic volume
per a gram of macromolecular substance at infinite
dilution.   The reduced viscosity which pertains to
solutions of finite concentration has the same units
of volume per gram of substance.
    Polyelectrolytes (i.e., dextran sulfate) in water
do not exhibit linear reduced viscosity curves over
the concentration range that macromolecular substances are usually studied (1%).   The reduced
viscosity curve is a continuously increasing function
with dilution (Fig. 2).   The continual increase with
dilution does not occur indefinitely.   At extremely
low concentrations (10^{-3}) the reduced viscosity
function decreases very rapidly with further dilution.
Should the diluting aqueous solvent contain an
electrolyte such as NaCl, etc., the reduced viscosity

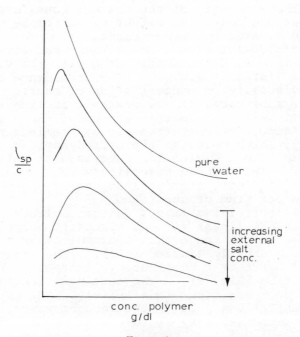

$$\frac{\eta}{\eta_o} = \eta_r \qquad \text{RELATIVE VISCOSITY} \qquad \frac{t}{t_o} = t_r$$

$$\frac{\eta}{\eta_o} - 1 = \eta_{sp} \qquad \text{SPECIFIC VISCOSITY} \qquad \frac{t}{t_o} - 1 = t_{sp}$$

$$\frac{\frac{\eta}{\eta_o} - 1}{c} = \frac{\eta_{sp}}{c} \qquad \text{REDUCED VISCOSITY} \qquad \frac{\frac{t}{t_o} - 1}{c} = \frac{t_{sp}}{c}$$

$$[\eta] \qquad \text{INTRINSIC VISCOSITY}$$

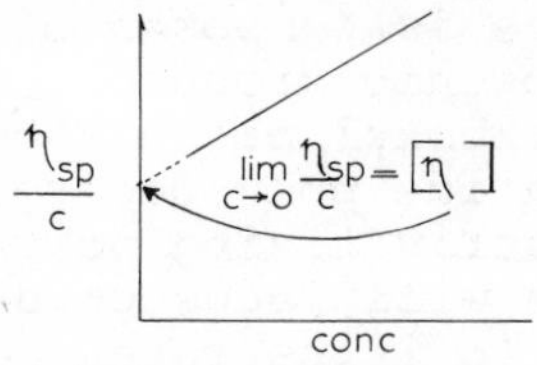

$$\frac{\eta_{sp}}{c} \qquad \lim_{c \to 0} \frac{\eta_{sp}}{c} = [\eta]$$

conc

UNITS OF $\frac{\eta_{sp}}{c}$ AND $[\eta]$ ARE dl/g or ml/g

Figure 1. Viscosity functions

$\frac{\eta_{sp}}{c}$

pure water

increasing external salt conc.

conc. polymer g/dl

Figure 2.

curves exhibit maxima at finite concentrations.   The
larger the external salt concentration, the smaller
the reduced viscosity values and the further to the
right the maximum reduced viscosity value tends to
appear (Fig. 2).   A linear dependence of reduced
viscosity on polyelectrolyte concentration is obtained
in the presence of a sufficiently high external salt
concentration.

The viscosity behaviour of polyelectrolytes is
governed by the first, second and third electroviscous
effect (2) (Fig.3).   The 1st electroviscous effect
arises because of the difference in size of the macro
ion and the counter ions.   In an hydrodynamic
gradient, the small counter ions are swept along more
rapidly than the much larger macro ion.   Charge
separation of the counter ion cloud from the macro ion
occurs.   Because the two are coupled by a coulombic
type interaction, the larger macro ion acts as a
brake on the counter ion movement.   This increases the
viscosity of the solution.   In solution, as the
liquid flows, macro ions will be driven past each
other because of the hydrodynamic gradient.   Should
the highly charged macro ions pass closely,coulombic
repulsive forces will come into play.   The faster
moving macro ion will deviate from its initial linear
pathway.   Again, excess energy is expended and the
viscosity of the medium is increased.   The larger
the charge on the macro ion, the stronger will be the
2nd electroviscous effect.   The 3rd electroviscous
effect arises because of the interaction of the
charges that are attached to the macromolecular backbone.   In the case of a flexible macromolecular coil,
this interaction expands the coil to an average
conformation which minimizes the repulsive interactions.
At the new equilibrium conformation (larger than that
of the neutral macromolecule), the contractile free
energy of the macromolecular backbone is equal to the
expansive coulombic free energy arising from ionization.   The increased macromolecular coil size
enhances the viscosity of the solution.   The viscosity
behaviour to the left of the maxima in Fig. 2 is
primarily due to the 2nd electroviscous effect, while
that to the right is primarily due to the 3rd electroviscous effect.

Not all of the counterions of a polyelectrolyte
are free to move about.   The free ions form a counte-
ion cloud about the polyion, whereas the immobilized
ions are bound to a specific site or point of the
macromolecular backbone.   This model was presented
earlier in the polyelectrolyte dissolution equation.

As the polyelectrolyte solution is diluted more and
more of the site bound counter ions are released.
This builds up the charge on the macro ion which
expands, which in turn increases the reduced viscosity.
Expansion on dilution, however, cannot occur indefinitely. When the concentration of the external
ions of the solution become equal to or greater than
that of the counterions of the polyelectrolyte, ionization of the polyelectrolyte ceases. Further
dilution decreases the reduced viscosity because
expansion of the coil has ceased and the charged
particles are placed further and further apart, causing
a reduction in the 2nd electroviscous effect. This
is the origin of the maxima of the reduced viscosity
curves.

Dextran Polyelectrolyte Behaviour.

A sufficiently large external salt concentration
will yield linear reduced viscosity-concentration
plots. Linearity, however, does not insure that the
viscosity behaviour is that of the neutral macromolecule. Fig. 4 shows the reduced viscosity
behaviour of a B-512 linear dextran($[\eta]$ = 0.164 dl/g)
and a branched dextran B-742($[\eta]$ = 0.158 dl/g) and
the sulfate derivatives derived from them. Despite
linearity, the reduced viscosities of the sulfates
are higher than those of the neutral molecules by a
factor of about two. The difficulty in collapsing
the sulfate macromolecular coil to the size of the
neutral macromolecule may stem from one of two factors
or a combination of both. Introduction of the sulfate
group may decrease the flexibility of the macromolecular backbone. A rigid backbone tends to produce
a more extended macromolecular conformation which
would exhibit higher reduced viscosities. Alternately,
although strong long range coulombic interactions
have been eliminated by the external salt, it may be
that short range interactions of the ion pairs exist.

Effect of Degree of Substitution. The reduced
viscosities of a number of potassium dextran sulfates
of differing degree of substitution derived from
B-742($[\eta]$=0.158) are shown in Fig. 5. Increasing
the degree of substitution enhances the reduced
viscosity and shifts the position at which the
maximum reduced viscosity appears to the left.
Increasing the number of ionogenic groups produces
more charge on the macro ion, causing greater
expansion of the coil. On dilution, further ionization

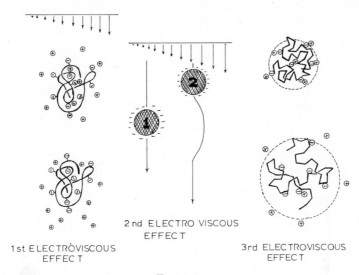

2 nd ELECTRO VISCOUS
EFFECT

1st ELECTROVISCOUS
EFFECT

3rd ELECTROVISCOUS
EFFECT

Figure 3.

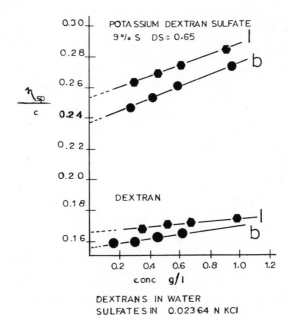

DEXTRANS IN WATER
SULFATES IN 0.023 64 N KCl

Figure 4.

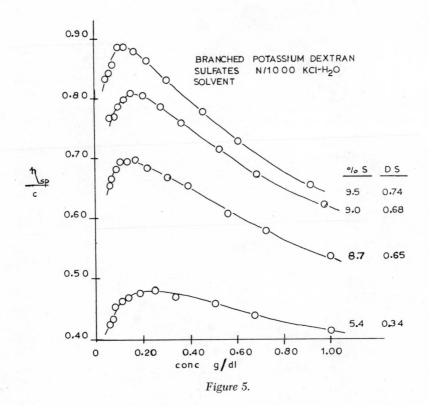

Figure 5.

and expansion occur in each case. The higher the
degree of substitution, the further must the polyelectrolyte solution be diluted to match the external
salt concentration with the counter ion concentration
of the polyelectrolyte. Similar viscosity behaviour
is observed for linear dextran sulfates and for
branched and linear carboxymethyl dextrans. The
typical polyelectrolyte viscosity curves exhibited by
dextran suggest that the macromolecular backbone is
fairly flexible and that the coil can undergo
expansion on acquiring charge.

Effect of Molecular Weight. Fig. 6 indicates the
effect of molecular weight on potassium carboxymethyl
dextran reduced viscosity curves. The degree of
substitution is constant and the molecular weight
varies from 73,000 to 135,000. The reduced viscosities
increase with molecular weight and the concentration
at which the reduced viscosity maximum appears is
identical for all three molecular weights. It would
appear that the molecular weight does not influence
the extent or degree of ionization and that the
expansion is directly proportional to the number of
substituted anhydroglucose units in the macromolecule
$[(\eta_{sp}/c)_{max}$ of 135,000 molecular weight sample
approximately 2x $(\eta_{sp}/c)_{max}$ of 73,000 molecular
weight sample$]$. This suggests that the interaction
of the ionogenic groups is a localized or nearest
neighbor interaction. Should it be otherwise, then
each charge of polyelectrolyte would interact with
every other, compounding the interactions. The
higher molecular weight macromolecule carrying more
charge would register a non-proportionate reduced
viscosity. The linear proportionality between
molecular weight and the maximum reduced viscosity
would not exist. To show more quantitatively that
the same ionization and expansion process is occurring
with the different molecular weights, the data of
Fig. 6 can be plotted in terms of a relative expansion
factor R_n vs the concentration of potassium carboxymethyl dextran as in Fig. 7. The numerator of R_n is
the maximum reduced viscosity and the denominator is
the reduced viscosity at a polyelectrolyte concentration greater than that at which the maximum viscosity
appears. The coincidence of the linear plots for
the three molecular weights indicates an ionization
expansion mechanism that is identical for the three
polyelectrolyte samples.

Effect of Macromolecular Structure. In Fig. 8 are

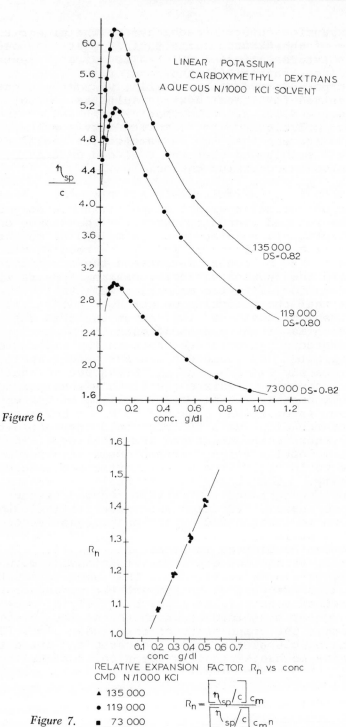

Figure 6.

Figure 7.

plotted the reduced viscosity curves for a branched
and a linear dextran sulfate of identical degree of
substitution derived from dextrans of near identical
intrinsic viscosity (0.16 dl/g). Although both
viscosity curves exhibit typical polyelectrolyte
characteristics, there are differences.   The
reduced viscosity maximum for the branched polyelectrolyte appears at a different concentration than that of
the linear.   In the more concentrated region, the
viscosity is higher for the branched dextran polyelectrolyte.   The linear dextran polyelectrolyte
the higher viscosity in the more dilute solutions.
A viscosity curve cross over has occurred.   Similar
behaviour occurs when the viscosity curves are run
in aqueous N/2000 KCl solution.   The maxima and the
cross over point shift to the left and the reduced
viscosities are larger (Fig. 9).   The difference
between the two dextrans rests in their structure.
The branched dextran sulfate contains some thirty
percent non 1,6 linkages, while the linear contains
something less than five percent.   Identical
intrinsic viscosities of the neutral macromolecules
dictate that a larger charge density exists in the
coils of the branched dextran sulfate.   This results
in a greater expansion for the branched polyelectrolyte
than for the linear.   As dilution occurs, the
branched species reaches its limit of expansion
earlier because of its   structural makeup.   The
linear macroion continues to expand in the absence of
structural limitation.   Hence the cross over.   The
same features are observed when the linear and
branched dextrans are converted to the carboxymethyl
derivative (Fig. 10).

        Effect of Nature of the Ionogenic Function.   In
Fig. 11 are plotted the reduced viscosity concentration
curves for a potassium carboxymethyl dextran of degree
of substitution of 0.21 and a potassium dextran
sulfate of degree of substitution of 0.34.   Obviously,
the higher the degree of substitution of ionogenic
group, the larger the reduced viscosity does not hold.
The potassium dextran sulfate of higher degree of
substitution exhibits a lower reduced viscosity.
Since the macromolecular backbone and the counter ion
are identical, the inversion observed can only be due
to the nature of the ionogenic group and its interaction
with the counter ion.   Potassium dextran sulfate
appears to ionize less than does potassium carboxymethyl dextran, allowing less expansion of the macro
ion.

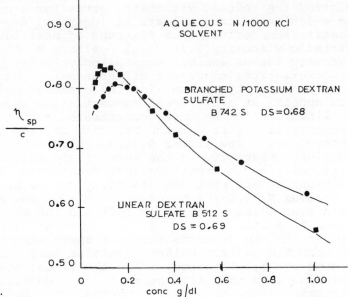

Figure 8.

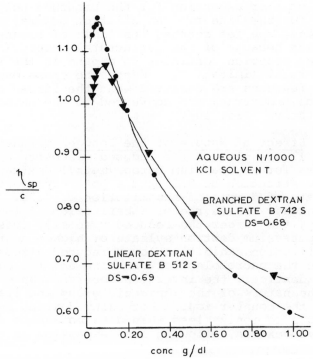

Figure 9.

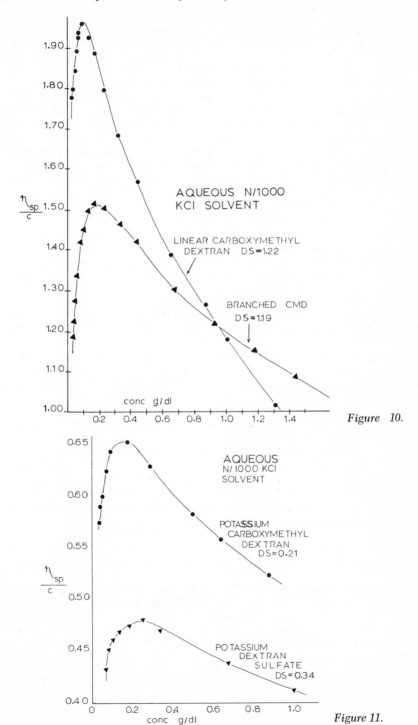

Figure   10.

Figure 11.

Effect of Counter Ion. Reduced viscosity curves
of dextran sulfates with Li, Na, K, and Cs as counter
ions are shown in Fig. 12. The degree of substitution
is 1.09 and the solvent is water. The macromolecular
backbone and the ionogenic group are identical for
the four polysaccharide polyelectrolytes. The only
difference between them is the counter ion. To
obtain this result, it must be that each alkali metal
cation interact to a different degree with the sulfate
function. Interestingly, a maximum reduced viscosity
is exhibited by the cesium dextran sulfate. This
would indicate that the cesium counter ion binds very
tightly and the extent of ionization of the salt is
much less compared with that of the other counter
ions. Similar reduced viscosity behaviour is
observed for Li, Na and K carboxymethyl dextran of
DS 0.84. A cesium carboxymethyl dextran reduced
viscosity curve was not obtained for this sequence.
Excellent linear correlation is obtained between the
reduced viscosity at a particular concentration and
the crystal radius of the ion (Fig. 13). Charge
density of the ion does not seem to be a very important
factor for if it were Li should be the most tightly
bound, yet it is not. Lithium ions are the most
hydrated of the alkali metal series. Cesium is the
least. This lack of hydration may be responsible
for the tight binding of the cesium ion.

Ion Binding. It is obvious from the preceding
discussion that the nature of the interaction of the
counter ion with the charged site on the macromolecular backbone is important in defining the polyelectrolyte solution properties. It is also obvious
that not all counter ions interact with the ionogenic
site of the polyion in an identical manner. At this
point in time, an accurate description of polyelectrolyte ion binding has not evolved. It is, however,
accepted that site binding and atmospheric binding
are the two modes of interaction. Site binding
gives rise to an ion pair and although a wide variety
of ion pairs can be defined and probably can exist in
a polyelectrolyte solution, for our purpose it is
fruitful to think of an ion pair as an entity in which
the counter ion does not have mobility. In atmospheric binding, the counter ion does have mobility
and on average there is a slightly higher concentration
of these mobile bounter ions in the vicinity of the
polyion than in the bulk of the solution.
Ion selective electrodes are responsive to free
or mobile cations and offer a means of detecting

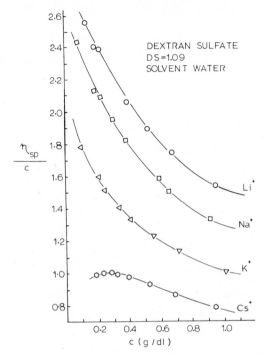

Figure 12.

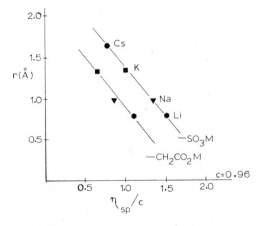

Figure 13.

TABLE I.   "IONIZATION CONSTANTS" OF SODIUM DEXTRAN
SULFATE OF DS 0.48.

Polyelectrolyte Conc. g/dl.	Percent Ionization	Ionization Constant $Kx10^3$
0.9600	39.7	5.7
0.7200	52.1	9.3
0.4800	47.0	4.4
0.3600	42.2	2.3
0.2400	47.3	1.7
0.0960	64.4	2.5
0.0480	66.1	1.4
0.0240	72.7	1.1
0.0120	81.5	1.0

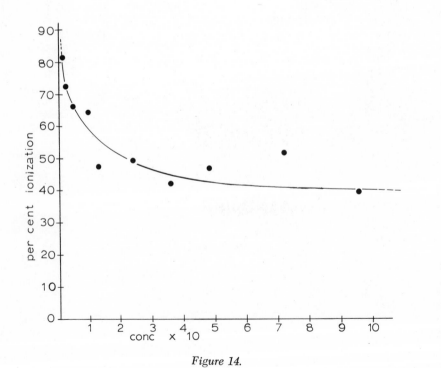

Figure 14.

their concentrations. If it is assumed that the sodium ion of an ion pair does not interact with the sodium ion electrode and that only the mobile ions do, then the percent ionization of the polyelectrolyte can be calculated. The percent ionization should increase on dilution of the polyelectrolyte if the dilution theory and the explanation of the reduced viscosity curves is correct in principle. The percent ionization for a sodium dextran sulfate of DS 0.48 has been plotted as a function of polyelectrolyte concentration in Fig. 14. It is seen that the percent ionization increases with dilution, paralleling the reduced viscosity curves of polyelectrolytes in water or in low external salt aqueous media. Interestingly, the curve of Fig. 14 is reminiscent of the percent ionization curve of acetic acid, suggesting that it might be possible to calculate an ionization constant for the sodium dextran sulfate. Values for the "ionization constant" of the sodium dextran sulfate at different polyelectrolyte concentrations are given in Table I. Allowing for the scatter in the raw data (Fig. 14), it would appear that within limits the value of K is constant. The raw data for the three most dilute solutions which do not appear to be scattered indicate more conclusively the constancy of K. In the event that the latter could be obtained for polyelectrolytes, then by analogy to acetic acid, where the counter ion (the proton) is certainly site bound, it can be argued that the sodium ion, or counter ion, is site bound. An ionization constant for polyelectrolytes would allow simple quantification of the interaction or ion pairing of counter ions with the macro ion.

Literature Cited.

1. Jeanes, A., J. Polymer Sci., Symposium Series (1974) 45, 209.

2. Conway, B.E. and Dobry-Duclau, A. in "Rheology", ed. Eirich, F.R., 3, 89, Academic Press, N.Y. (1960).

11

Some Rheological Properties of Gum Solutions

JOHN H. ELLIOTT

Research Center, Hercules Inc., Wilmington, DE 19899

End-use applications of water-soluble polymers, including extracellular microbial polysaccharides, are almost exclusively based upon the rheological properties which they confer upon the final system. A rather detailed knowledge of the rheological behavior of aqueous solutions of such polymers is essential for selection of the most suitable gum for a given end use. This paper will review some general rheological properties of aqueous gum solutions, including suitable experimental instrumentation. Supermolecular structure may be present in certain gum solutions, which gives rise to time dependent rheological behavior. Finally the use of rheological data in selecting gums for specific end uses will be illustrated.

Rheological Background

In this paper, we shall be concerned primarily with data obtained in viscometric or simple shear flows(1). Here there is a non-zero velocity component in only one direction in the medium. Familiar examples are the flows in capillary, concentric cylinder, and cone and plate instruments. The simplest case is that of the Newtonian liquid, where the shear stress, S (dynes/cm.^{2}) is directly proportional to the shear rate, $\dot\gamma$ (sec.$^{-1}$); the constant of proportionality being the viscosity, η (poise),

$$S = \eta\dot\gamma \qquad (1)$$

Here, the viscosity is a constant independent of shear rate. Gum solutions show this behavior at high dilu-

tions. As the concentration (or molecular weight) is
increased to the point where entanglement occurs(2),
however, this situation no longer prevails and the
viscosity becomes a function of the shear rate,
decreasing with increasing shear rate. This is called
pseudoplasticity or shear thinning. The complete
pseudoplastic curve is shown in Figure 1. It should
be emphasized that within the time scale of conventional laboratory measurements, pseudoplasticity is a
reversible phenomenon. There are three principal
regions of this log η vs. log $\dot{\gamma}$ curve. At very low
shear rates, the viscosity is Newtonian. This zero
shear or first Newtonian viscosity, η_0, is a function
of the molecular weight, M, and concentration, C, of
the polymer. It has been found that a number of
polymer-solvent systems follow the relationship(3)

$$\eta_0 \propto C^5 M^{3.4} \qquad\qquad (2)$$

The variation of η_0 with $M^{3.4}$ has been well established for polymer melts(3).

  As the shear rate is increased, a decrease in
viscosity is observed. After a relatively short
transition region, log η becomes linear in log $\dot{\gamma}$.
This is the so-called power law region and may cover
many decades in shear rate. This is generally described by the following equations

$$S = K(\dot{\gamma})^n \quad \text{and} \qquad\qquad (3a)$$

$$\eta = K(\dot{\gamma})^{n-1} \qquad\qquad (3b)$$

The slope of the log η vs. log $\dot{\gamma}$ line in this region
is n-1. If n is one, the liquid is Newtonian; if n
is less than one, it is pseudoplastic; if n is greater
than one, the system is dilatant or shear thickening.
This behavior is generally observed in systems containing a high volume fraction of solids. The power
law was considered as an empirical relationship for
many years; however, Scott-Blair(4) has given a simple
theoretical derivation based on the concept of the
breaking of "linkages" by shear.

  As the shear rate is increased another transition
zone is observed, followed by a Newtonian region, the
infinite shear or second Newtonian viscosity, η_{∞}.
This region is observed at very high shear rates and
is very difficult to study experimentally. The value
of η_{∞} appears to show only slight dependence on
molecular weight, in contrast to η_0 and may be orders
of magnitude lower than η_0(5). The η_{∞} region is of
little practical importance and is rarely observed.

The following is a qualitative rationalization of
the general pseudoplastic curve shown in Figure 1(6).
In the η_0 region, the time scale of the measurement is
sufficiently long that structure or entanglements are
not disrupted and Newtonian flow is observed. As the
shear rate is increased, the time scale becomes
shorter and the polymeric units cannot relax. Structure is broken down and the polymer molecules tend to
become oriented in the flow direction. These effects
increase with increasing shear rate, giving rise to
power law behavior. As the shear rate is increased to
very high values, breakdown and orientation have gone
as far as possible and a further increase in shear
rate does not affect them. The flow is then
Newtonian, the η_{00} region.

The η_0 and power law regions are most important
in characterizing aqueous gum systems. Some generali-
zations can be made about behavior in these regions.
Figure 2 shows log η vs. log $\dot\gamma$ plots for xanthan gum
in aqueous solution. At the higher concentrations,
2500 and 1500 ppm., the η_0 region lies at shear rates
below 10^{-2} sec.$^{-1}$. This region is quite apparent,
however, at the lower concentrations, 500 and 250 ppm.
It is also apparent that as the polymer concentration
is increased, the transition from η_0 to non-Newtonian
flow occurs at lower values of the shear rate. The
power law slope at higher concentrations is quite
steep and not very sensitive to concentration. As
concentration is reduced, this slope becomes less
steep, and, while not shown in Figure 2, at very low
concentrations Newtonian behavior is observed.

Solutions of polymers, having a long-chain
branched structure, will show a lower η_0 than a linear
polymer of the same weight average molecular weight.
This has been extensively studied in the case of
linear and long-chain branched polyethylenes. A
classic study in this field is that of Busse and
Longworth(7).

The effect of salts on the rheological properties
of aqueous gum solutions is a matter of considerable
practical importance. The presence of salts markedly
lowers the viscosity of dilute solutions of polyelectrolytes; in fact, we have observed a decrease of
over three decades in the η_0 value of a 2500 ppm.
solution of a polyacrylamide, having anionic functionality, in going from distilled water to a 2.2% brine.
This effect may be largely attributed to the polyelectrolyte effect, that is, the ionic strength of the
medium reduces the repulsion between adjacent charges
on the polymer chain. This results in a conforma-

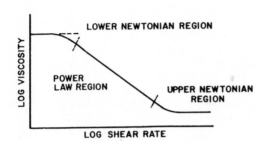

Figure 1. Viscosity as a function of shear rate for a pseudoplastic or shear thinning system

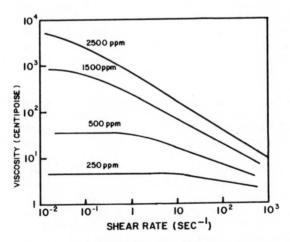

Figure 2. Viscosity of xanthan solutions as a function of shear rate and of concentration

tional change from an extended configuration toward
that of a random coil.

In general, salt solutions are poorer solvents
for water-soluble polysaccharides than distilled water.
As a consequence, the viscosities of dilute solutions
of these polymers and their intrinsic viscosities are
lower in salt solutions than in pure water. It is
frequently observed, however, that concentrated solutions show higher viscosities when salts are present.
Tager(8) has carried out extensive studies of the
viscosities of concentrated solutions of organic
soluble polymers in good and poor solvents. In the
case of polar polymers, viscosities in poor solvents
may be several decades higher than those in good solvents. This is a consequence of the formation of
relatively strong supermolecular structures by the
polymer molecules in the poorer solvent. The same
considerations are applicable to water-soluble gums.

It is apparent from the earlier discussion that a
realistic rheological characterization of gum solutions
requires the determination of its viscosity as a function of shear rate over at least several decades of
shear rate. Concentric cylinder or cone and plate
rheometers, covering a wide range of shear rates, are
the most suitable instruments. In our own work, when
the viscosity-shear rate curve was needed over more
than six decades of shear rate, it was necessary to
use three different instruments, the Weissenberg
rheogoniometer, the Haake Rotovisco, and the Hercules
Hi-Shear Viscometer, to cover the low, intermediate
and high shear rate ranges, respectively. Excellent
agreement between instruments was found in the regions
of overlap.

Gum solutions show elastic as well as viscous
properties. These are readily determined by imposing
a sinusoidal strain of small amplitude upon the
sample, for example, using a cone and plate instrument.
The resulting stress wave is sinusoidal and has the
same frequency as the imposed strain wave(9). It is,
however, out of phase with the strain wave, the phase
angle, δ, lying between $0°$ and $90°$. This may be
resolved into a component in phase with the strain,
from which the dynamic modulus, G', may be calculated.
The stress wave component in quadriture with the
imposed strain yields the dynamic viscosity, η'. In
and near the η_0 region, η' and the viscosity in steady
shear are in good agreement when the steady shear is
plotted against $\dot{\gamma}$ and η' against the frequency, ω, in
radians/second, i.e., considering ω in oscillation
equal to $\dot{\gamma}$ in steady shear(10). At higher frequen-

cies η· generally is lower than the corresponding
steady shear viscosity.

Figure 3 shows η, η', and G' for a 2% solution of
sodium carboxymethylcellulose (CMC) in water. It is
seen that η' lies somewhat below η. G' increases with
frequency while η' decreases, which is the expected
behavior.

Supermolecular structure, which is often present
in gum solutions, gives rise to a variety of rheological phenomena. In contrast with pseudoplastic
behavior, these effects are time dependent. The most
common is thixotropy, which has been defined as a
reversible gel-sol transition. It is observed experimentally as a decrease in viscosity with time at a
constant shear rate. Eventually an equilibrium viscosity value is reached. If shearing is stopped, the
viscosity will rise to its original value, as the
structure in the system reforms. A different and
widely used method of characterizing thixotropy,
developed some years ago by Green and Weltman(11),
involves programmed increases in shear rate from rest
to a high value (the up curve), followed by a rapid
decrease back to zero shear rate (the down curve).
A typical example of such a shear rate-shear stress
curve is shown in Figure 4. The area of the loop is a
measure of the work per unit volume per second for
thixotropic breakdown, under the conditions of the
experiment. The extrapolated intercept of the down
curve on the shear axis can be considered as a yield
stress. It must be emphasized that if the programmed
rate of increase in shear rate used in obtaining the
up curve is changed, the area of the hysteresis loop
and the value of the extrapolated yield stress will,
in general, be different.

The concept of a yield stress is very useful in
the rheological characterization of systems having
supermolecular structure. It was proposed some time
ago by Bingham(12). Whether or not it really exists
has been debated in the intervening years. Bingham's
original definition is

$$(S - \sigma_0) = \eta \dot{\gamma} \tag{4}$$

where the shear stress, S, must exceed the yield value,
σ_0, before flow can occur. This is shown in Figure 5.
In practice, this type of behavior is never strictly
observed. The experimental flow curve does not inter-
sect the abscissa sharply but curves in toward the
origin, as shown by the dotted line. A straight portion of the curve, at higher shear rates, however, may
be extrapolated to the abscissa to give a value of σ_0.

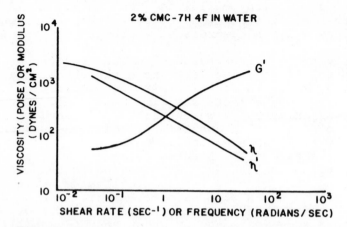

Figure 3. Dynamic viscosity and modulus and steady shear viscosity as functions of frequency or shear rate for 2.0% CMC in water

Figure 4. Hysteresis loop treatment. Stress vs. shear rate for a thixotropic material—arrows indicate increasing and decreasing shear rate

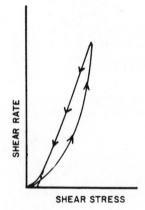

Perhaps what may be operationally considered as a
yield stress is merely the presence of a relaxation
time which is very much greater than the time scale of
the experimental measurement.

There is a fundamental difference between steady
shear and dynamic measurements in the case of systems
exhibiting a time dependent rheological response. The
total strain, to which the sample is subjected in
steady shear, is determined by the shear rate and the
time that it is applied. The total strain can thus be
very large, leading to extensive structure breakdown
and polymer orientation. Dynamic measurements, on the
other hand, are carried out at low strain amplitudes.
Under these conditions, there is little or no structure breakdown or polymer orientation, and the properties of the sample are measured essentially at rest.
Dynamic and steady shear measurements supplement each
other to give a more complete rheological characterization. It must be remembered, however, that the
rheological state of the system may be quite different
in the two types of measurements.

Rheological Properties in End Uses

Jeanes(13) has recently published an extensive
review on the applications of extracellular microbial
polysaccharide-polyelectrolytes. In this paper,
rheological properties in end uses will be illustrated
by several examples with which the writer has had
first-hand experience. These, unfortunately, have not
been involved primarily with extracellular microbial
polysaccharides; however, they do illustrate the
application of rheological characterization to end
uses.

Food Systems

Sodium carboxymethylcellulose, often called
cellulose gum or CMC, is a widely used component of
food systems. It may act as a suspending agent,
thickener, protective colloid, humectant, and to control the crystallization of some other component. CMC
is classified by the Food and Drug Administration
under "substances that are generally recognized as
safe" (Gras). CMC is prepared by the reaction of
alkali cellulose with sodium chloroacetate and is a
polyelectrolyte. Important parameters in characterizing CMC are the average degree of polymerization
(DP), the average number of anhydroglucose units per
molecule; and the average degree of substitution (DS),

the average number of carboxymethyl groups per anhydroglucose unit.

It was early recognized in our rheological characterization of CMC solutions and gels that samples having the same nominal chemical composition and solution viscosity could show markedly different rheological properties. Earlier work showed that CMC prepared under conditions giving more uniform substitution gave pseudoplastic solutions. If these conditions were not followed, thixotropic solutions resulted. This is particularly true for the lower DS levels. Our work led to the conclusion that a very small quantity of unsubstituted crystalline cellulose residues, existing as fringe micelles, act as cross-linking centers and enable a three-dimensional network to be formed(14,15).

Nijhoff was granted a U.S. patent(16) on the use of CMC, having low DS, to form unctuous gels for low calorie spreads. This prompted a study of the rheological properties of unctuous materials. It was found that when such materials (e.g., butter, mayonnaise and ointments) were subjected to an imposed sinusoidal strain, which was greater than the linear viscoelastic limit, the resulting stress wave was not sinusoidal, but in many cases approached a square wave(17). Similar observations have been reported by Komatsu, et al.(18). Our results were interpreted in terms of a modified Bingham body, consisting of an elastic, a frictional and a viscous element connected in series. The response of this model to steady shear and to imposed sinusoidal shear has been calculated(19), and the model has proved to be useful in characterizing structured systems(20). Komatsu et al. interpreted their experimental results using the Casson equation, which will be discussed later. Figure 6 shows the effect of DS on the steady shear properties of 5% CMC solutions and gels. Curve A for the sample having a DS of 0.7 is typical of a viscoelastic system, and this was confirmed by dynamic measurements. Curve B for a sample of DS 0.4 shows a steeper rise in the stress and also a stress overshoot. The Curve C for an experimental sample having a DS of 0.18 shows the very sharp peak and rapid stress decay, characteristic of unctuous systems. Curve D shows dynamic measurements on the DS 0.18 sample. The square nature of the stress curve is obvious.

Table I illustrates the types of CMC used in a number of food products. The connection between the type used in a given system and the solution rheological properties discussed above will be apparent.

The yield stress, as a useful rheological concept, was discussed earlier. It is frequently diffi-

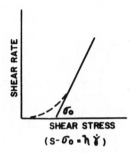

Figure 5. *Stress vs. shear rate for a Bingham body*

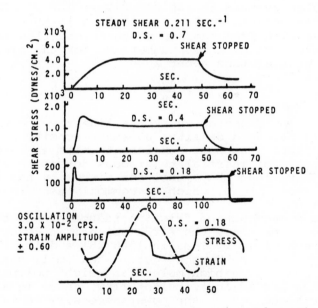

Figure 6. *Five percent CMC in water; effect of degree of substitution on stress response*

cult to establish even semi-quantitatively, using
Bingham's relationship (equation 4). The Casson
equation,

$$S^{\frac{1}{2}} - \sigma_o^{\frac{1}{2}} = A\ \dot{\gamma}^{\frac{1}{2}} \qquad (5)$$

which was derived for suspensions, has proved to be
very useful in determining small values of σ_o from a
plot of $S^{\frac{1}{2}}$ against $\dot{\gamma}^{\frac{1}{2}}$(24). As an example, chocolate
milk is formulated with about 0.03% κ-carrageenan
which prevents settling of the cocoa. The κ-
carrageenan causes a considerable increase in the viscosity of the milk-sugar system and the existence of a
yield stress has been postulated. Plots of S vs. $\dot{\gamma}$
could not be extrapolated to give a value of σ_o;
however, the Casson plots, shown in Figure 7, are
linear for values of $\dot{\gamma}^{\frac{1}{2}}$ between zero and one and permit a reliable extrapolation.

Friction Reduction

     Rheological behavior, discussed so far, has been
confined to systems in laminar flow. The phenomenon
of drag reduction is observed only in turbulent flow.
The transition from laminar to turbulent flow in a pipe
occurs when the Reynolds Number, R, for the system
becomes greater than about 2000

$$R = \frac{dv\rho}{\mu} \qquad (6)$$

d is the pipe diameter, v is the linear velocity of
the fluid having a density, ρ, and viscosity μ. R is
dimensionless when self consistent units are used, and
is the ratio of inertial to viscous forces in the
fluid. Note that the symbol for viscosity has been
changed from η to μ, which is commonly used in
engineering.
     The phenomenon of friction reduction has been
known for some time, the first scientific description
having been given by Toms in 1948(25) and it is often
referred to as the Toms effect. When small quantities
(10-500 ppm.) of a high molecular weight polymer are
added to a liquid in turbulent flow, there is a
dramatic reduction in the power necessary to maintain
the same flow rate. When the same concentration of
polymer is added to the liquid in laminar flow, the
only effect observed is a slight increase in viscosity,
which may be scarcely detectable. If pressure drop is
measured along a tube, the fluid velocity or Reynolds

TABLE I

SOME USES OF CMC IN FOODS

FOOD PRODUCT	PROPERTY CONFERRED BY CMC	CMC TYPE DP*)	DS	S**)	REFERENCE
BAKED GOODS	WATER RETENTION, CONTROL OF BATTER VISCOSITY	H,M	0.7	NO	21, 22, 23
DOUGHNUTS	GREASE HOLDOUT	H,M	0.7	NO	21, 22, 23
STARCH SYSTEMS, WHIPPED TOPPINGS	INHIBITION OF SYNERESIS	H,M	0.7	NO	21, 22
SYRUPS, BEVERAGES, JUICES	THICKENER, VISCOSIFIER	H,M	0.7	NO	21, 22, 23
ICE CREAM	TEXTURE, BODY, CONTROL OF SUGAR AND ICE CRYSTALLIZATION	H,M	0.7	NO	21, 22, 23
PET FOODS (SEMIMOIST)	BINDER	H,M	0.7	NO	21, 22, 23
CONFECTIONS	CONTROL OF SUGAR CRYSTALLIZATION	L	0.7	NO	21, 22, 23
LOW CALORIE SYRUPS	VERY SMOOTH TEXTURE	H	0.7	YES	21, 22, 23
LOW CALORIE SPREADS	UNCTUOUSNESS	H,M	0.4	NO	16, 17

*) H,M,L, HIGH, MEDIUM, OR LOW VISCOSITY   **) S, UNIFORM SUBSTITUTION

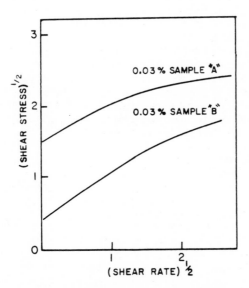

Figure 7.   Casson plots for κ-carrageenan–milk–sugar system

number being held constant, the percent friction
reduction, % FR, is given by

$$\%FR = 100 \left(\frac{\Delta P_o - \Delta P}{\Delta P_o} \right) = 100 \left(1 - \frac{\Delta P}{\Delta P_o} \right) \qquad (7)$$

where ΔP_o is the pressure drop for the pure liquid and
ΔP is the pressure drop for the same liquid containing
a low concentration of polymer. Percent friction
reductions as high as 70-80% have been observed. The
importance of this effect to those industries where
large quantities of liquids must be pumped is obvious.
Extensive research studies have also been carried by
United States and foreign naval laboratories(26,27).
Many water-soluble polymers, including cellulosics and
microbial polysaccharides have been tested as friction
reduction additives(28). The mechanism by which these
polymers produce friction reduction is not clearly
understood. In general, high molecular weight and a
linear structure give a more efficient polymer. The
diameter of the test section is a very important
parameter; greater friction reductions usually are
observed in smaller diameter tubes. It is known that
these additives thicken the viscous sub-layer at the
pipe wall. Viscoelastic effects almost certainly play
an effect and it may be that elastic energy storage by
the polymer molecule interacts with the small, energy
dissipating, turbulent eddies. Polymer supermolecular
structure may also play a role and it has been suggested that this could be a significant factor in the
diameter effect(29). The shear in turbulent flow can
degrade the polymer molecules. Guar gum and sodium
carboxymethylcellulose are more shear resistant but
less effective friction reducers than poly-(ethylene
oxide) and poly-(acrylamides)(27). The field of
friction reduction is very active, and the following
reviews are recommended(24,30,31).

Flow Through Porous Media

        This is another situation in which the flow is
not laminar. The basic equation for flow through
porous media is Darcy's law:

$$\frac{Q}{A} = \frac{k}{\mu} \frac{\Delta P}{\Delta l} \qquad (8)$$

Q is the flow rate in cm.^{3}/sec. through a cross
sectional area of A (cm.^{2}), $\frac{\Delta P}{\Delta l}$ is the pressure

gradient in atmospheres/cm., μ is the viscosity in
centipoise, and k is the permeability in darcies.   In
cgs units, permeability has the units cm.^{2} and one
darcy equals 9.87×10^{-9} cm.^{2}.   The darcy is the
commonly used unit of permeability in geophysical
work.   The ratio k/μ is the mobility.   This is a most
important parameter in porous media studies, and will
be discussed in terms of the use of polymer solutions
as mobility buffers in enhanced oil recovery.   The
economics of enhanced oil recovery techniques are now
feasible, in view of the energy shortage and the high
cost of imported petroleum.

The importance of the mobility in such an operation arises from the fact that if one liquid, e.g.,
oil, is to be pushed out by another immiscible liquid,
e.g., water or a polymer solution, the latter must
have an equal or lower mobility to prevent fingering
or water breakthrough, which would bypass recoverable
oil in the formation(32).   Consideration of Darcy's
equation indicates two mechanisms by which addition
of a polymer to water can lower the mobility.   It can
increase the viscosity and/or it can lower the permeability of the porous medium to the aqueous solution.
The latter is brought about by polymer adsorption or
entrapment.   In general, both mechanisms appear to be
operating, however, one or the other tends to
predominate.

The flow of a polymer solution through a porous
medium is not a laminar or viscometric flow and
unusual effects may be observed.   For example,
poly(acrylamide) solutions show pseudoplastic behavior
in viscometric flows.   In porous media, however, such
solutions appear to exhibit dilatant properties(33).
Indeed a poly(acrylamide) solution in water showed
both pseudoplastic and dilatant responses, depending
upon the flow rate(34).

As a polymer molecule moves through the pores of
a porous medium, it is subjected to accelerations and
decelerations.   These, together with the stretching
deformation which occurs as it passes through a fine
pore, introduce elastic and relaxation effects which
are absent in viscometric flows.   Thus the flow
behavior of polymer solutions in a porous medium
cannot be predicted from viscometric measurements,
but, in general must be determined in the specific
porous medium of interest.

Xanthan gum is being currently tested for use in
mobility control for enhanced oil recovery.   Available
information indicates that it operates to lower
mobility primarily by increasing viscosity.   As shown

in Figure 2, xanthan solutions show pseudoplastic behavior down to quite low concentrations. This is desirable, giving low viscosities at the high shear rates encountered during injection, but significantly higher viscosities when moving through the oil bearing formation, where shear rates are in the range of 0.1 to 10 sec.$^{-1}$.

Literature Cited

(1) Middleman, S., "The Flow of High Polymers", p. 8, Interscience, New York, 1968.
(2) Peterlin, A., "Non-Newtonian Viscosity and the Macromolecule", p. 225, Advances in Macromolecular Chemistry, Volume 1, W. M. Pasika, Ed., Academic Press, New York, 1968.
(3) Reference 1, p. 172.
(4) Scott-Blair, G. W., Rheol. Acta (1965) 4, 53.
(5) Reference 1, p. 101.
(6) Lenk, R. S., "Plastics Rheology", p. 12, Wiley Interscience, New York, 1968.
(7) Busse, W. F., and Longworth, R., J. Polymer Sci. (1962) 58, 49.
(8) Tager, A. A., Rheol. Acta (1974) 13, 831.
(9) Ferry, J. D., "Viscoelastic Properties of Polymers", p. 12, Wiley, New York, 1970.
(10) Bueche, F., "Physical Properties of Polymers", p. 220, Interscience, New York, 1962.
(11) Green, H., and Weltman, R. N., Ind. Eng. Chem., Anal. Ed. (1943) 15, 201. Also Green, H., "Industrial Rheology and Rheological Structures", Wiley, New York, 1949.
(12) Bingham, E. C., "Fluidity and Plasticity", p. 217, McGraw-Hill, New York, 1922.
(13) Jeanes, A., J. Polymer Sci. (1974) Symposium No. 45, 209.
(14) Ott, E., and Elliott, J. H., Makromol. Chem. (1956) 18/19, 352.
(15) deButts, E. H., Hudy, J. A., and Elliott, J. H., Ind. Eng. Chem. (1957) 49, 94.
(16) Nijhoff, G. J. J., U.S. Patent 3,418,133.
(17) Elliott, J. H., and Ganz, A. J., J. Texture Studies (1971) 2, 220.
(18) Komatsu, H., Mitsui, T., and Onogi, S., Trans. Soc. Rheol. (1973) 17:2, 351
(19) Elliott, J. H., and Green, C. E., J. Texture Studies (1972) 3, 194.
(20) Elliott, J. H., and Ganz, A. J., Rheol. Acta (1974) 13, 1178.

(21) Ganz, A. J., Food Product Develop. (1969) <u>3</u> (<u>6</u>), 65.

(22) Batdorf, J. B., "Industrial Gums", R.L. Whistler, Ed., Academic Press, New York, 1959.

(23) Glicksman, M., "Gum Technology in the Food Industry", Academic Press, New York, 1969.

(24) Scott-Blair, G. W., Rheol. Acta (1966) <u>5</u>, 184.

(25) Toms, B. A., "Proceedings International Rheological Congress, Holland 1948", p. II-135, North Holland Publishing Co., Amsterdam, 1949.

(26) Little, R. C., Hansen, R. J., Hunston, D. L., Kim, O. K., Patterson, R. L., and Ting, R. Y., Ind. Eng. Chem., Fundam. (1975) <u>14</u>, 283.

(27) Van der Meulen, J. H. J., Appl. Sci. Res. (1974) <u>29</u>, 161.

(28) Hoyt, J. W., Polymer Letters (1971) <u>9</u>, 851.

(29) Elliott, J. H., and Stow, F. S. Jr., J. Appl. Polym. Sci. (1971) <u>15</u>, 2743.

(30) Gadd, G. E., "Encyclopedia of Polymer Science and Technology", Vol. 15, H. F. Mark, Chairman Editorial Board, N. M. Bikales, Executive Editor, Interscience-Wiley, New York, 1971.

(31) Lumley, J. L., J. Polym. Sci. (1973) Macromolecular Reviews, <u>7</u>, 263.

(32) Collins, R. E., "Flow of Fluids Through Porous Materials", p. 196, Reinhold, New York, 1961.

(33) Burcik, E. J., Producers Monthly (1967) <u>31, No. 3</u> 27.

(34) Jones, W. M., and Davies, O. H., Nature Phys. Sci. (Nov. 13, 1972) <u>240</u>, 46.

12

Rheology of Xanthan Gum Solutions

P. J. WHITCOMB
General Mills Chemicals, Inc., 2010 E. Hennepin Ave., Minneapolis, MN 55413

B. J. EK and C. W. MACOSKO
Dept. of Chemical Engineering and Material Science, University of Minnesota, Minneapolis, MN 55455

Xanthan gum is a high molecular weight polysaccharide produced by fermentation with the bacterium
"Xanthomonas campestris". Adding ½% of this biopolymer
increases water's viscosity by a factor of 100,000 at
low shear rates; yet at high shear rates, the factor is
reduced to 10. This remarkable shear thinning ability
(pseudoplasticity) can be used to great advantage. In
fact the main use of xanthan gum is rheology control.
In the past most data on xanthan gum rheology has
been taken over a limited shear rate range and is
relative, not absolute. Figure 1 presents typical
shear stress vs. shear rate data. Only two decades of
shear rate are covered and the shear rate is given as
rpm, relative units. The use of arithmetic scales for
this plot make it difficult to resolve solution properties. Figure 2 presents a typical viscosity vs.
shear rate plot. The use of log scales aids in interpretation of the flow curves. However, shear rate
covers only two decades and is again reported as rpm.
While relative data is useful for comparisons
under specified conditions, there are many advantages
to having absolute data, where the units have physical
significance. Absolute data is independent of the
instrument or geometry used to gather the data.
Relative data is not. This means that a very broad
shear rate range can be covered by compiling results,
obtained in absolute units, from several instruments.
Shear rate overlap between instruments insures the
integrity of the data, since a systematic error due to
a particular instrument will be detected. With absolute data, constitutive or empirical relationships can
be used to model the dependence of viscosity on shear
rate. Such models are essential to predict flow

160

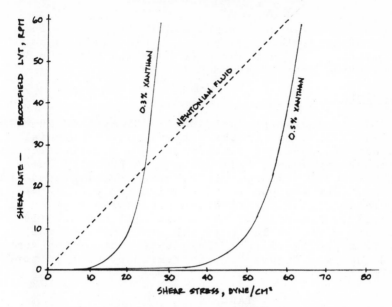

Figure 1. Xanthan gum pseudoplasticity

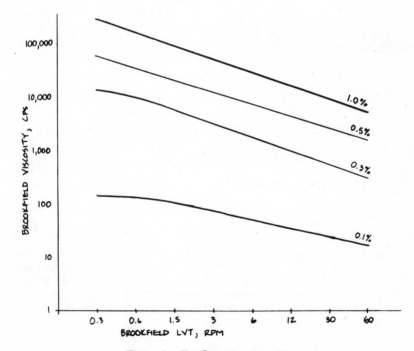

Figure 2. Xanthan gum viscosities

behavior, design equipment, and interpret rheology in
terms of molecular structure. It is the purpose of
this work to collect absolute data and evaluate it in
terms of an appropriate model.

Description

    A commercial product, GALAXY®XB Xanthan gum (lot
D5353A), was used for this study. The commercial
product was purified further using a modification of a
procedure by Jeanes (1): The gum is hydrated in a
water-ethanol mixture, centrifuged, precipitated and
washed. The wet gum is then dried and ground. A
schematic of the purification procedure is given in
Figure 3. Analysis of the commercial and purified
gums are given in Table I. The viscosity of the
purified gum is slightly less than that of the commercial gum, see Figure 4. The remainder of this paper
deals only with the purified gum.

TABLE I

ANALYSIS

	Commercial Product	Purified Gum
Solution O.D. 400 NM	.19	.07
Water - Wt. %	10.6	13.0
Nitrogen - Wt. % MFB	0.67	0.69
Protein - Wt. % MFB	3.8	3.9
Ash - Wt. % MFB	13.8	7.9
Sodium - Wt. % MFB	4.4	1.9
Phosphorus - Wt. % MFB	0.32	0.23

    All solutions used in this study were prepared
from gum hydrated in distilled water. The solutions
were prepared by sprinkling the ground xanthan gum
onto the sides of a vortex formed in a high speed
blender. All gum concentrations are reported on a
moisture free basis.

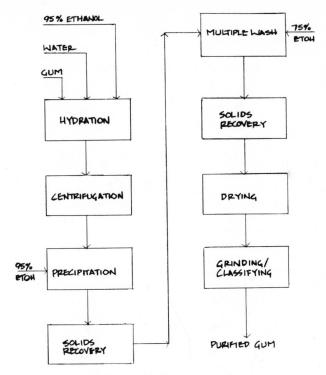

Figure 3.   Purification procedure

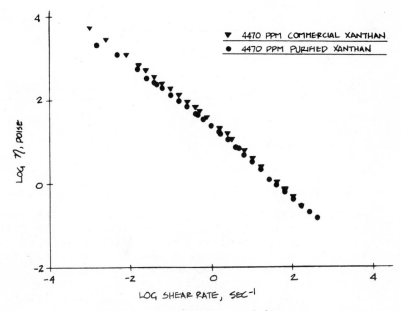

Figure 4.   Commercial vs. purified viscosity

Intrinsic viscosity was determined using a
Cannon-Ubbelohde capillary dilution viscometer, number
598 - size 100. All other rheological data was taken
on either a Weisenberg Rheogoniometer (2) or a Rheo-
metrics Mechanical Spectrometer (3) using cone and
plate fixtures. The cone and plate geometry with its
associated equations is pictured in Figure 5. The
primary advantage of this geometry is that shear stress
and shear rate are nearly constant throughout the
sample. This means that graphical differentiation of
the data is not needed to obtain the shear stress vs.
shear rate curve in absolute units (4). The instruments used provide steady rotational speeds which can
be varied to cover a wide range of shear rate $\dot{\gamma}$. By
varying cone angles the range of each instrument was
expanded. The shear stress τ_{12} was calculated from
the measured torque on the stationary member of the
cone and plate. The shear rate dependent viscosity,
$\eta(\dot{\gamma})$, was determined for each shear rate. The combinations of cones and instruments used are shown in
Table II. Note how the shear rate ranges overlap.

TABLE II

INSTRUMENT RANGES

INSTRUMENT	CONE ANGLE RAD. B	CONE RADIUS CM. R	SHEAR RATE SEC^{-1} $\dot{\gamma}$	
			MIN.	MAX.
WEISSENBERG	0.1	3.6	5.95×10^{-4}	148
	0.06	3.6	9.92×10^{-4}	247
MECHANICAL SPECTROMETER	0.1	3.6	10^{-2}	400
	0.04	3.6	0.25	10^{3}
	0.01	1.25	10^{-1}	4000

Every cone and instrument combination was calibrated with a Newtonian oil, dioctyl phthalate. All data
was taken at room temperature, which varied between
25°C and 29°C. The xanthan data was not corrected for
temperature. Figure 6 presents the calibration data
corrected to 25°C. Observe the approximate range
covered by the Rheogoniometer and Mechanical Spectro-

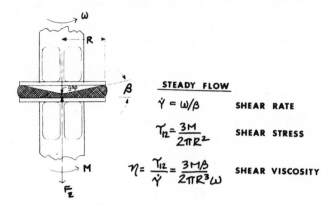

$$\text{STEADY FLOW}$$

$$\dot{\gamma} = \omega/\beta \qquad \text{SHEAR RATE}$$

$$\tau_{12} = \frac{3M}{2\pi R^2} \qquad \text{SHEAR STRESS}$$

$$\eta = \frac{\tau_{12}}{\dot{\gamma}} = \frac{3M\beta}{2\pi R^3 \omega} \qquad \text{SHEAR VISCOSITY}$$

Figure 5.   Flow between a cone and plate

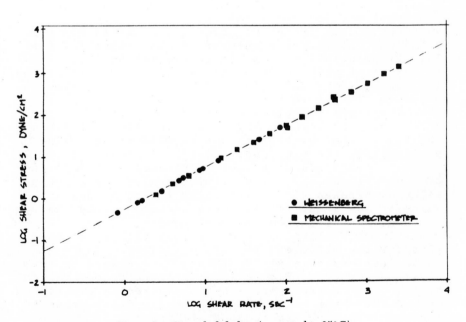

Figure 6.   Dioctyl phthalate (corrected to 25°C)

meter. The following equation was used for temperature
correction (5)

$$\log \eta_{25} = \log \eta_T + \frac{d \log \eta}{dT} (\Delta T)$$

Where η_{25} - Viscosity corrected to 25°C
       η_T - Viscosity at T°C
       ΔT - (T-25)°C
       $\frac{d \log \eta}{dT}$ - Slope of log η vs. T between 25°C
                         and T°C.

Data

        Data was taken over a wide shear rate range for
each of four concentrations of the purified gum;
10,000, 2,000, 1,000 and 447 ppm. This data was first
plotted as shear stress vs. shear rate then as viscosity vs. shear rate. In the graphs of shear stress
the dashed line represents the shear stress plot of
water. In the graphs of viscosity the base line at
10^{-2} poise is the viscosity of water. All graphs are
in absolute units.
        The shear stress plots of the 10,000 and 2,000
ppm solutions are shown in Figure 7. Notice six
decades of shear rate are covered for the 10,000 ppm
solution. At very low shear rates the curve for the
10,000 ppm solution seems to be flattening out. It
appears that this solution has a definite yield stress,
τ_y, near 13 dyne/cm^2. The 2,000 ppm solution shows no
evidence of a yield stress.
        The viscosity plots of the 10,000 and 2,000 ppm
solutions are shown in Figure 8. The yield stress of
the 10,000 ppm solution is not as evident on a viscosity plot. However, Newtonian regions are more
readily identified on a plot of viscosity than on one
of shear stress. The viscosity curve of the 2,000 ppm
solution appears to be flattening toward a constant
value at low shear rates. This Newtonian value is
known as the zero shear rate viscosity, η_0.
        Figure 9 shows the shear stress vs. shear rate
plot for the 1,000 and 447 ppm solutions. There are
no indications of a yield stress. The viscosity vs.
shear rate curves, Figure 10, for these concentrations
appear to be approaching a Newtonian region at the low
end of the shear rate range covered. These concentrations are expected to go Newtonian at low shear
rates, as did the 2,000 ppm solution.
        The intrinsic viscosity of the purified gum was
determined from a plot of inherent viscosity

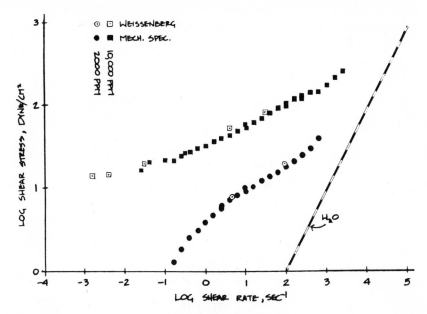

Figure 7.   Xanthan gum rheology

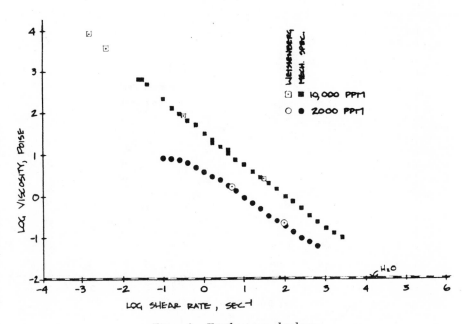

Figure 8.   Xanthan gum rheology

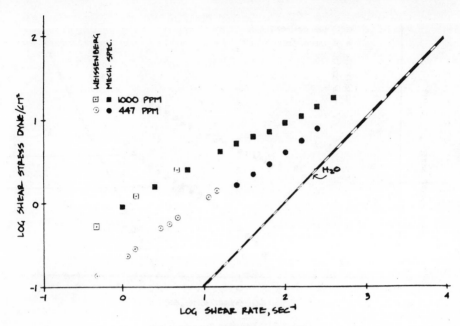

Figure 9.   *Xanthan gum rheology*

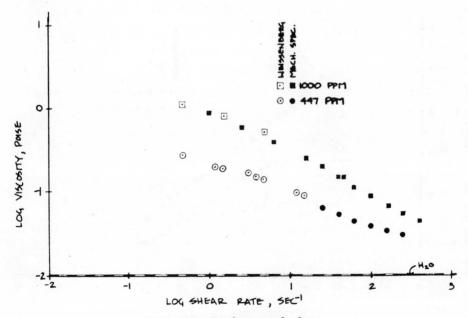

Figure 10.   *Xanthan gum rheology*

extrapolated to zero concentration. This determination
is illustrated in Figure 11 and yields an intrinsic
viscosity of 35.70 deciliters/gm.

Discussion and Conclusions

There appears to be a critical concentration
between 2,000 and 10,000 ppm above which a yield stress
exists. Identifying and measuring yield stress is
important for predicting the long term stability of
suspensions. Concentrations below the critical value
will have a Newtonian region at very low shear rates.
This region is characterized by the zero shear rate
viscosity, η_0. All concentrations should have a
Newtonian region at very high shear rates. Without
this region the solution viscosities will go below
that of their solvent, water. This region is characterized by the infinite shear rate viscosity, η_∞.
Many mathematical expressions have been proposed
to model the pseudoplastic behavior exhibited by
xanthan. The most widely used is the power law of
Ostwald (6).

$$\eta = K \, \dot{\gamma}^{n-1}$$

The power law is appealing because of its simplicity,
there are only two adjustable parameters n and K. K,
the viscosity at 1 sec $^{-1}$, measures consistency and n,
the flow index, measures pseudoplasticity. Many
empirical and analytical solutions for complex flows
have been worked out using the power law. Examples
being laminar and turbulent pipe flow, (7) annular
flow (8), flow through porous media (9), mixing
characteristics and heat transfer problems (10) to cite
a few.
The main disadvantage of the power law is its
failure in the regions of very low shear, η_0 or τ_y, and
very high shear, η_∞. More sophisticated models can
account for the Newtonian regions in high and low shear
rate regions. However, for the xanthan concentrations
studied the log data is linear for several decades of
shear rate. Table III shows the power law fit of our
xanthan data. The high determination index and the
broad range of shear rate fit for each concentration
verify the utility of this model.

TABLE III

POWERLAW CONSTANTS

CONC. PPM	K (DYNE·SEC) /CM^2	n DIMENSIONLESS	$\dot{\gamma}$ RANGE, SEC^{-1} MIN.	MAX.	DETERMINATION INDEX
10,000	35	.23	2.5×10^{-2}	$2.5 \times 10^{+3}$	.98
2,000	3.4	.39	1.0×10^{-1}	$6.25 \times 10^{+2}$	.97
1,000	0.98	.49	4.71×10^{-1}	$4.0 \times 10^{+2}$	.99
447	0.23	.64	4.71×10^{-1}	$2.5 \times 10^{+2}$	.99

Examination of Table III reveals K increases with increasing concentration while n decreases. This means that the higher the concentration of xanthan the thicker the solution. However, higher concentrations are more pseudoplastic, have a lower n value, so at high shear rates all concentrations, at least those of 10,000 ppm and less, approach the viscosity of water. The power law fit for the 10,000 and 1,000 ppm solutions are shown in Figure 12. Note the difference in K and n for the two concentrations, see Table III. Power law data is very valuable in evaluating solution properties and solving practical problems.

Another type of theory has been used to explore intrinsic viscosity. Assuming the conformation of a xanthan molecule can be approximated by a cylindrical rod, it is possible to estimate its characteristic length. Using the theory of Khalik and Bird (11) rod length can be determined by the following expression:

$$L^3 = \frac{[\eta]\ (45)\ (MW)\ (\ln\ (L/D)\ )}{2\pi\ N}$$

Where
L — Rod length
[η] — Intrinsic viscosity
MW — Molecular weight
D — Rod diameter
N — Avogadro number

This expression can be solved by iteration if rod length is the only unknown. The molecular weight of xanthan has estimated to be in the range of 1.4 to 3.6 x 10^6 (Dintzis et al (12)). The intrinsic viscosity was measured to be 35.7 deciliters/gm. We have

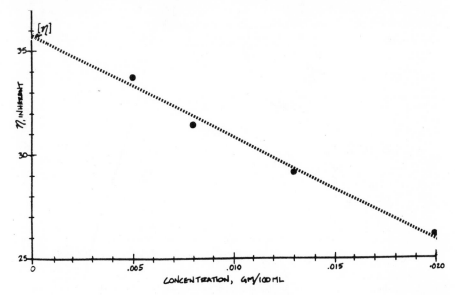

Figure 11. Intrinsic viscosity

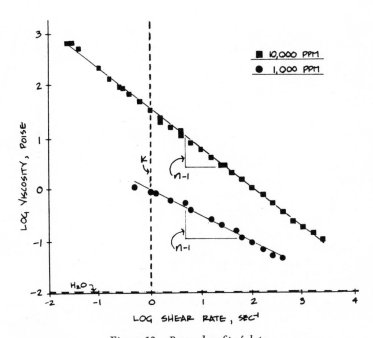

Figure 12. Power law fit of data

estimated the diameter of a xanthan "rod" to be in the range of 16 to 40Å, the length calculation is quite insensitive to diameter,

TABLE IV

CALCULATED LENGTH

MOLECULAR WEIGHT GM/MOLE	DIAMETER Å	[η] ML/GM	LENGTH MICRONS
1.4×10^6	16	3570	0.73
1.4×10^6	40	3570	0.68
3.6×10^6	16	3570	1.01
3.6×10^6	40	3570	0.96

Table IV gives the results of the rod length calculation. It would appear that a xanthan "rod" has a length between 0.7 and 1.0 microns. This is in good agreement with Holzwarth's (13) membrane chromatography measurements. He found that essentially all xanthan particles can pass through a membrane with 1.0 micron pores but are blocked by a membrane with 0.8 micron pores.

Nomenclature

D - Rod diameter, (microns)
K - Power law constant, intercept
L - Rod length (microns)
M - Torque (gm-cm)
MW- Molecular Weight
n - Power law constant, slope
N - Avogadro number
O.D. Optical Density
R - Cone radius (cm)
T - Temperature (°C)
β - Cone angle (radians)
γ̇ - Shear rate (sec^{-1})
η - Viscosity (poise)
$η_0$- Zero shear rate viscosity (poise)
$η_\infty$- Infinite shear rate viscosity (poise)
[η] Intrinsic viscosity (deciliters/gm)

τ_{12} Shear stress (dyne/cm^2)
τ_y Yield stress (dyne/cm^2)
ω^y Angular speed (radians/sec.)

Literature Cited

(1)   Jeanes, A., Pittsley, J.E,, Senti, F.R,,
        J. Applied Polymer Sci, (1961), 5,
        p. 519-526,
(2)   Van Wazer, J.R,, et al, "Viscosity and Flow
        Measurement", p. 113-116, Wiley, N.Y.1963.
(3)   Macosko, C.W., Starita, J., S.P.E. Journal,(1971),
        27, p. 38-42.
(4)   Middleman, S., "The Flow of High Polymers", p.28,
        Wiley, New York, 1968.
(5)   Willey, S.J. Ph.D. Thesis, University of
        Minnesota, 1976.
(6)   Ostwald, W., Kolloid-Zeitschrift, (1925), 36,
        p. 99-117.
(7)   Metzner, A.B., Reed, J.C., A.I.Ch.E. J., (1955),
        1, p. 434-440.
(8)   Mishra, P., Mishra, I., A.I.Ch.E. J.,(1976), 22,
        p. 617-619.
(9)   Sheffield, R.E., Metzner, A.B., A.I.Ch.E. J.,
        (1976), 22, p. 736-744.
(10) Wilkinson, W.L., "Non-Newtonian Fluids",
        Pergamon, New York, 1960.
(11) Abdel-Khalik, S.K., Bird, R.B., Biopolymers,
        (1975), 14, p. 1915-1932.
(12) Dintzis, E.R., Babcock, G.E,, Tobin, R.,
        Carbohydrate Research. (1970), 13,
        p. 257-267.
(13) Holzwarth, G., "Polysaccharide from Xanthomonas
        Campestris: Rheology, Solution Conforma-
        tion, And Flow Through of Petroleum
        Chemistry, A.C.S., New York Meeting,
        April 4-9, 1976.

13

Synergistic Xanthan Gels

I. C. M. DEA and E. R. MORRIS

Unilever Research, Colworth/Welwyn Laboratory,
Sharnbrook, Bedford, MK44, 1LQ, Great Britain

Although showing considerable evidence of strong intermolecular interactions in solution, xanthan on its own does not
gel. On mixing with another non-gelling polysaccharide, locust-
bean gum, however, firm rubbery gels are formed at low polymer
levels, typically around 1% total polysaccharide, with most
effective xanthan utilisation at locust-bean gum: xanthan ratios
of about 3:1 (1,2,3). The molecular origin of the synergism has
until recently (4,5,6) remained obscure.

In this paper we will attempt to answer two questions:-
1) What is the mechanism of formation of these mixed gels?
2) Why should two polysaccharides of such diverse origins
interact:

Technological Relevance of Biological Function

Many of the industrial uses of polysaccharides rest solely
on their water binding capacity and high viscosity at low concentrations. Increasingly, however, more sophisticated applications
depend on detailed molecular structure, and exploit in vitro the
specific function of the polysaccharide in vivo. This is
particularly true in gelling systems. Thus agar, carrageenan,
furcellaran, pectin and alginate all have a structural role in
nature, which gives rise directly to their gelling behaviour.

Alginate, for example, is the major structural polysaccharide
of Brown Seaweed. Chemically it is a block co-polymer of
D-mannuronic and L-guluronic acid, in which the homopolymeric
polyguluronate sequences are capable of forming very strong intermolecular cross-links, while polymannuronate or alternating
sequences show far less tendency to associate (5-11). The
relative amount of the various block types is under enzymic control at the polymer level (12), providing subtle biological control of the mechanical properties of different parts of the plant
at different stages of maturation. These structural variations
are reflected directly in the gelation properties of the poly-

174

saccharide in vitro. Thus alginate extracted from growing
fronds giving appreciably less rigid gels than material from
mature stipes.
    Knowledge of the natural role of industrially important
polysaccharides can therefore provide valuable insight into their
effective commercial utilisation. Thus any understanding we can
gain of the biological utility of xanthan synergism may well be
of interest from a technological as well as an academic standpoint.

Polysaccharide Gel Structure

    An understanding of the mechanism of synergistic gelation is
perhaps best approached from the principles established for
single polysaccharide systems (5,7,13). The gel state is a meta-
stable half-way house between the solid state, with molecules in
regular ordered conformations packed together with little hydration, and the solution state, with extensively hydrated polymer
molecules in random conformations. The structural integrity of
polysaccharide gels is maintained by intermolecular association
into long, structurally regular junction zones, in which the
molecules adopt the same ordered conformation as in the solid
state. These junction zones are therefore essentially crystalline, although they may only involve two polymer chains, and are
terminated typically by an interruption in the regular covalent
structure (e.g. in alginate the occurrence of a mannuronate
residue would terminate association of polyguluronate sequences).
Such junctions are held together by a regular array of non-
covalent intermolecular bonds, whose energy offsets the loss of
conformational entropy in forming such a rigid assembly, and
whose co-operative action elevates the lifetime of the junctions
to a macroscopic timescale.
    The junction zones are then linked by regions of the molecule
which are structurally incapable of forming stable associations,
or are prevented from doing so by network constraints. These
non-associated regions presumably maintain essentially the same
disordered conformation as in solution, and solubilise the gel
network by extensive hydration. Thus both associating and non-
associating molecular regions are essential for gelation, too
much association leading to precipitation, and too little
preventing formation of a cohesive network.

Aggregation of Rigid Structures

    In gels of carrageenan or agar the primary mechanism of
intermolecular association is by double helix formation.
Further development of the gel network, however, involves
association of helices into larger aggregates (14-16). The
extent of aggregation increases as electrostatic repulsion between the molecules decreases, being greatest for the neutral

agarose helix. Aggregation of rigid rod-like species is a highly
favourable process, since, unlike the formation of ordered
junctions between flexible molecules, little loss of conformational energy is involved. Moreover, above a critical concentration,
alignment of extended structures becomes a geometrical neccessity
(17) providing an additional drive to aggregation.

Similar considerations apply to the association of polysaccharide chains which, while not totally rigid, have severely
restricted mobility about the glycosidic linkages between
adjacent residues, and therefore tend to favour extended conformations close to that found in the solid state. Locust-bean gum
falls into this category. Chemically it is a galactomannan, with
a β 1-4 linked mannan backbone and α 1-6 linked galactose substituents, which occur in long blocks, as shown in Figure 1 (18).
Fractions of varying galactose content may be obtained from
commercial locust-bean gum samples by utilising the greater
solubility of the more highly substituted chains (4). As outlined in Figure 2, the solid state conformation of the mannan
chain is an extended, two-fold, ribbon-like structure, virtually
identical to that of cellulose, since both have a 1-4 diequa-
torially linked hexopyranose backbone, and differ only in the
orientation of O(2).

Under normal conditions there is no evidence of aggregation
of galactomannan molecules in solution. Freezing and thawing
concentrated locust-bean gum solutions, however, yields stable
gels whose gel strength increases with decreasing galactose
content. On freezing, the formation of ice crystals must pro-
gressively raise the concentration of polymer in the remaining
unfrozen solution, to the point where alignment of the chains
becomes sterically essential, until finally the chains pack
together as in the solid state. Once formed the chain-chain contacts between the 'smooth' unsubstituted mannan backbone regions
appear to be sufficiently energetically favourable to hold the
molecules together on thawing. In the resultant gel network, the
substituted 'hairy' regions presumably act as the solubilising
interconnecting regions which prevent precipitation of the
associated chains.

Xanthan is not alone in showing synergism with galactomannans, but shares this property with both agar and carrageenan.
For both of these it has been established that the mechanism of
gelation involves association of unsubstituted backbone regions
of the galactomannan, in an ordered conformation, with the rigid,
ordered, helical structure of the polysaccharide (19). We must
therefore consider whether a similar mechanism operates for
xanthan - galactomannan interactions.

Xanthan Native Conformation

An essential requirement of this model is a rigid, regular,
ordered structure with which the mannan chain can align. Until

recently no such rigid conformation was suspected for xanthan. Spectroscopic and rheological studies now show, however, that the native conformation of the molecule is a rigid rod, which is only melted out to the expected random coil under conditions of very high temperature and low ionic strength (4,6,20,2f). The order-disorder transition is conveniently monitored as a sharp sigmoidal discontinuity in a single-wavelength optical rotation (Figure 3).

Detail of the ordered structure is still the subject of X-ray studies (22), but it is known to involve the charged tri-saccharide side-chains aligning with the cellulose backbone of the xanthan molecule, in a 5-fold helical structure. The solution properties of xanthan are entirely consistent with extensive orientation and aggregation of the rigid molecular rods (20-21), analogous to the previously discussed aggregation of agarose and carrageenan double helices.

Molecular Origin of Xanthan Synergism

The strength of interaction between xanthan and galacto-mannans is closely correlated with the degree of substitution of the mannan chain. Guar gum, in which the ratio of mannose to galactose is close to 2:1 does not gel with xanthan in any con-centration, although a slight viscous interaction is observed (1). Soft, fairly weak gels are obtained with gum tara, where the mannose to galactose ratio is around 3:1, as against 4:1 in locust-bean gum, which gives far stronger and more rigid gels. Even greater enhancement of gel properties is found for hot water soluble locust-bean gum fractions in which the ratio is 5:1 or more. Thus, once more, unsubstituted regions of the mannan back-bone are implicated in junction formation.

Optical rotation studies of synergistic xanthan gelation provide strong evidence that the native ordered xanthan conforma-tion is present in the mixed gel. As shown in Figure 4, the characteristic sigmoidal curve which accompanies the order-disorder transition persists in the presence of galactomannan, and is essentially complete before the onset of gelation. This interpretation is confirmed by X-ray studies on oriented films prepared from synergistic xanthan gels (23), which show virtually the same diffraction features as for xanthan alone. We therefore conclude that xanthan – galactomannan gels are crosslinked by co-operative association of unsubstituted mannan regions in a regular ribbon-like conformation, to the native ordered xanthan structure, as outlined schematically in Figure 5.

Biological Utility

Xanthan is the extracellular polysaccharide from Xanthomonas campestris, which causes blight in cabbage crops. Other related Xanthomonas species are parasitic upon a wide variety of other plants, and all appear to synthesise ordered polysaccharides

Figure 1.   Schematic for a galactomannan such as locust bean gum.   (○) 1→4 linked β-D-mannopyranose residues; (●) α-D-galactopyranose residues.

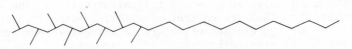

Figure 2.   Schematic for galactomannan conformation

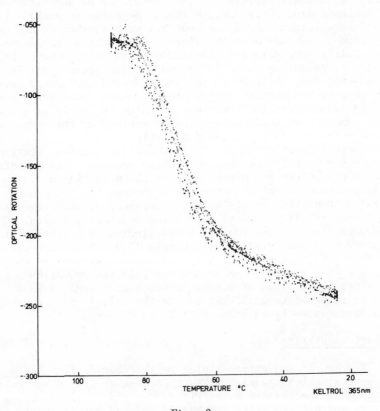

Figure 3.

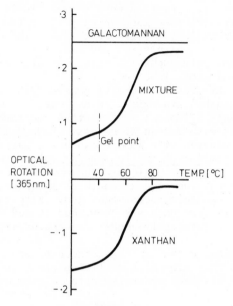

Figure 4.

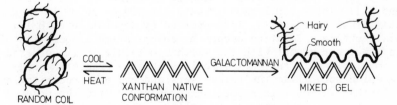

Figure 5

capable of synergistic interaction (24). Xanthan is a complex
molecule (25,26), and a large number of steps are involved in its
biosynthesis (27). It would be surprising (28,29) if the amount
of genetic information involved were stored simply to discharge
the relatively trivial functions which have so far been proposed
for extracellular bacterial polysaccharides. The existence of an
ordered native xanthan structure, and its affinity for specific
sequences in plant polysaccharide molecules also argue for a
sophisticated biological function.

To explore this further we have examined the scope and
specificity of xanthan synergism. Enhancement of microcrystalline cellulose gels (7) shows that xanthan can interact with
glucan as well as with mannan sequences. Indeed a very strong
interaction is observed with konjac mannan, a partially acetylated
polysaccharide from Amorphophallus konjac tubers, whose β 1-4
linked linear contains both glucose and mannose residues. This
material shows about the same strength of interaction with the
agar double helix as does locust-bean gum. With xanthan, however,
its interaction is very much stronger, and indeed recognisable
gels are formed at total polysaccharide concentrations as low as
0.05%.

There is evidence that at this concentration xanthan can
bind directly to the cell walls of living plant tissue (21,30,31).
It therefore appears that synergism with galactomannans may be
a co-incidental by-product of a natural role which involves
interaction with cellulosic materials on the cell wall surfaces
of the host plant. Such interactions may perhaps be involved in
recognition of appropriate sites for eventual colonisation by the
bacteria, or in preparation of the cell surface for attachment of
the parasite.

Abstract

Although neither xanthan nor locust-bean gum will gel alone
under normal conditions, mixed gels can be formed at total polysaccharide concentrations well below 1%. Chemically locust-bean
gum is a galactomannan, with a mannose to galactose ratio of
around 4:1. The α 1,6 linked galactose residues occur in long
blocks, ('hairy regions'), interspersed by unsubstituted β 1,4
mannan backbone. Gelation occurs by co-operative association of
these 'smooth regions' with the xanthan molecule in its ordered
conformation, while the 'hairy regions' act as connecting segments which solubilise the gel network and prevent precipitation.

Xanthan shares this synergistic behaviour with the ordered
conformations of carrageenan, furcellaran, and agar, but is
unique in showing a marked preference for interaction with β 1,4
glucose containing polysaccharides (including derivatised cellulose) rather than mannan. This suggests a possible biological
role for the polymer in substrate recognition by the synthesising
bacterium, Xanthomonas campestris.

Literature Cited

1. "Xanthan Gum, a Natural Biopolysaccharide for Scientific Water Control", Kelco Co., San Diego, California, 1972.
2. Rocks, J.K. Food Technol. (1971). 25, 476–483.
3. Federal Register, U.S. Government Printing Office, Washington D.C., Xanthan Gum, Section 121.1224 5376–5377, March 19th. 1969.
4. Dea, I.C.M. & Morrison, A. Advan. Carbohyd. Chem. Biochem. (1975). 31, 241–312.
5. Rees, D.A. Biochem. J. (1972). 126, 257–273.
6. Morris, E.R. in "Molecular Structure and Function of Food Carbohydrate". (Eds. Birch, G.G. & Green, L.F.) 125–130, Applied Science Publishers Ltd., London. 1973.
7. Rees, D.A. Advan. Carbohyd. Chem. Biochem. (1969). 24, 267–332.
8. Morris, E.R., Rees, D.A. & Thom, D. J. Chem. Soc. Chem. Commun. (1973). p. 245.
9. Grant, G.T., Morris, E.R., Rees, D.A., Smith, P.J.C. & Thom, D. FEBS Lett. (1973). 32, 195–197.
10. Morris, E.R., Rees, D.A., Sanderson, G.R. & Thom, D. J. Chem. Soc. Perkin II. (1975). pps. 1418–1425.
11. Morris, E.R., Rees, D.A. & Thom, D. In preparation.
12. Madgwick, J., Haug, A. & Larson, B. Acta Chem. Scand. (1973). 27, 3592–3594.
13. Rees, D.A. in "Biochemistry of Carbohydrates". (Ed. Whelan, W.J.) 1–42, Butterworths, London, 1975.
14. Rees, D.A., Steele, I.W. & Williamson, F.B. J. Polymer Sci. (C). (1969). 28, 261–276.
15. McKinnon, A.A., Rees, D.A. & Williamson, F.B., J. Chem. Soc. Chem. Commun. (1969). pps. 701–702.
16. Arnott, S., Fulmer, A., Scott, W.E., Dea, I.C.M., Moorhouse, R. & Rees, D.A. J. Mol. Biol. (1974). 90, 269–284.
17. Flory, P.J. Proc. Roy. Soc. ser. A. (1956). 234, 50–73.
18. Baker, C.W. & Whistler, R.L. (1975). Carbohyd. Res. 45, 237–243.
19. Dea, I.C.M., McKinnon, A.A., & Rees, D.A. (1972). J. Mol. Biol. 68, 153–172.
20. Morris, E.R. This Symposium.
21. Morris, E.R., Rees, D.A., Young, G., Walkinshaw, M. & Darke, A. J. Mol. Biol. Submitted.
22. Moorhouse, R. This Symposium.
23. Moorhouse, R. Personal Communication.
24. Schuppner, H.R. Jr. Australian Patent, 401,434. (1966).
25. Jansson, P.E., Kenne, L. & Lindberg, B. Carbohyd. Res. (1975). 45, 275–282.
26. Melton, L.D., Mindt, L., Rees, D.A. & Sanderson, G.R. Carbohyd. Res. (1976). 46, 245–257.

27. Sutherland, I.W.   This Symposium.
28. Sutherland, I.W.   Advan. Microbiol. Physiol.   (1972).
    8, 143–213.
29. Sutherland, I.W. & Norval, M.   Biochem. J.   (1970).   120,
    567–576.
30. Leach, J.G., Lilly, V.G., Wilson, H.A. & Purvis, M.R. Jr.
    Phytopathology.   (1975).   47, 113–120.
31. Lesley, S.M. & Hochster, R.M.   Canad. J. Physiol.   (1959),
    37, 513–529.

14

Xanthan Gum—Acetolysis as a Tool for the Elucidation of Structure

C. J. LAWSON and K. C. SYMES

Tate and Lyle Ltd., Group Research and Development, P.O. Box 68, Reading, U.K.

The development of microbial gums is now moving at an ever increasing pace and it appears likely that in the forseeable future a whole range of products will be available with properties not only reflecting and improving upon those found in many plant gums, but also of a novel nature to be exploited in as yet undeveloped applications. The most successful microbial gum to date is undoubtedly xanthan gum produced by Xanthomonas campestris, and this polymer is now commanding a market of several thousand tons per annum. The market position for xanthan gum has been developed through the unique physical properties which it shows, which are exploited for example in oil recovery and food applications. These properties are briefly, high viscosity, extreme pseudoplasticity stability to extremes of pH, salt tolerance and synergistic gelation in the presence of locust bean gum. The above properties are of course dictated by the primary, secondary and tertiary structures of the gum and it is necessary to determine these if any real understanding of the relationship between function and structure is to be obtained.

Early reports on the structure of xanthan gum, presented the repeating unit as being made up of glucose, glucuronic acid, mannose and the substituents pyruvate and acetate, in a 14 or 16 residue repeating unit. (1) (2) A repeating unit as large as this is unusual as most microbial gums have tri, tetra or pentasaccharide repeats. Also some of the chemical evidence was somewhat ambiguous, for example the assignment of the pyruvate as being linked to a glucose residue when it could equally have been associated with mannose. More recently two papers have been published, revising the structure and proposing a new pentasaccharide repeating unit containing the same sugar residues as before. We now provide further supporting evidence for the revised structure and suggest an approach to a rapid and convenient qualitative analysis of aspects of covalent structure of this and similar polysaccharides. The interest of Tate and Lyle in microbial gums was originally connected only with microbial alginate, (3) but as a natural consequence of involvement with gums generally, it was decided to examine

the possibilities of developing other microbial polysaccharides for incorporation into a possible range of products. One candidate for examination was
Xanthomonas campestris as the properties shown by xanthan gum were considered to be complementary to microbial alginate.

Traditionally, three basic lines of approach can be adopted in the
elucidation of polysaccharide structure. These are methylation analysis,
periodate oxidation and isolation of fragments which can be characterised;
the last approach only being of use when the polysaccharides have a repeating unit. The first two approaches had been reported in the previous papers
and therefore the third was the logical choice. Acetolysis was used because
aqueous acid hydrolysis often gives acidic oligomers from uronic acid containing polysaccharides and these are more difficult to characterise than
neutral fragments. Also at the time it was considered that acetolysis might
give a complementary result to partial acid hydrolysis which was also under
consideration. Acetolysis is, practically, a relatively straightforward
process performed at room temperature in approximately two days using
commonly available reagents. A sample of xanthan fermentation broth
obtained in batch culture of NRRL B1459 was taken. Purified xanthan gum
was recovered after bacterial cells were removed using high speed centrifugation and trypsin digestion, by alcohol precipitation.

The carefully dried gum was shaken with the acetolysis mixture of
reagents used by Morgan and O'Neill, (4) in studies on desulphated   λ -
carrageenan. The acetolysate was then poured into water, the acetylated
products extracted into chloroform, and deacetylated using methanolic
sodium methoxide in the usual way. The pale yellow syrup obtained in
high yield revealed, on chromatographic examination, a number of spots
in addition to the expected monosaccharides. The product was then
resolved into acidic and neutral fractions by separation on ion exchange
resin in the acetate form. (5) As hoped, the major proportion of oligo-
meric material was in the neutral fraction. The oligomers B to E were then
obtained in a purified state from the neutral fraction by a combination of
cellulose column and thick paper chromatography. (Figure 1) (Figure 2)

At this point in the work it was learned from Professor Rees of results
obtained by his group (6) and of Professor Lindbergs group (7) proposing
the revised structure of xanthan gum which has been mentioned earlier.
The revised repeating unit is based upon a cellulosic backbone with trisaccharide side chains occurring on alternate glucose residues. Analysis
of the acetolysis oligomers was therefore continued in order to ascertain,
whether they were consistent with the above structure.

The structure of oligosaccharide C will be used as an example of the
approach adopted, and the other oligosaccharides will only be mentioned
for the purpose of mentioning specific points of difference in their analysis.
This oligosaccharide was found to consist of glucose and mannose in a 2:1
ratio after hydrolysis and glc of the derived alditol acetates. This was

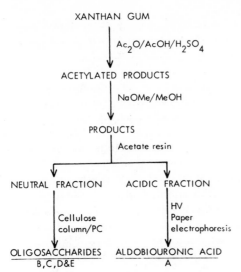

XANTHAN GUM

| $Ac_2O/AcOH/H_2SO_4$

ACETYLATED PRODUCTS

| NaOMe/MeOH

PRODUCTS

Acetate resin

NEUTRAL FRACTION   ACIDIC FRACTION

Cellulose   HV
column/PC   Paper
                     electrophoresis

OLIGOSACCHARIDES   ALDOBIOURONIC ACID
B,C,D&E   A

Figure 1. The acetolysis of xanthan gum

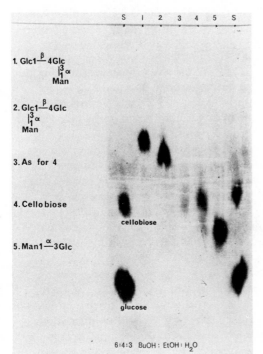

1. Glc1$\overset{\beta}{-}$4Glc
   $|\overset{3}{\underset{1}{}}\alpha$
   Man

2. Glc1$\overset{\beta}{-}$4Glc
   $|\overset{3}{\underset{1}{}}\alpha$
   Man

3. As for 4

4. Cellobiose

   cellobiose

5. Man1$\overset{\alpha}{-}$3Glc

   glucose

6:4:3 BuOH : EtOH : H_2O

Figure 2. Xanthan acetolysate neutral oligosaccharides

confirmed by a colorimetric assay of the ratio of carbohydrates to glucose which also showed that 50% glucose was lost on reduction with borohydride. The oligosaccharide was therefore shown to be trisaccharide having glucose as the reducing moiety. The mass spectrum of the T.M.S. ether of reduced C had ions at 451 as expected for fission of the substituted terminal hexose and 525 from the alditol moiety, thus providing evidence that the trisaccharide is not branched. Further more the series of fragments obtained at m/e ratios 103, 205 and 307 were those predicted from a 4-linked reduced glucose residue and this was confirmed with a deuterium labelling experiment. (Figure 3) The reduced oligosaccharide was then converted into the partially methylated alditol acetates of its component sugars which were analysed by gas chromatography. The retention times of the three resulting peaks were compared with literature values and in this way the substitution pattern of the central hexose in the trisaccharide was established as either 2-substituted mannose or 3-substituted glucose. (Figure 4)

The sequence of sugar residues in the trisaccharide and their anomeric configuration was then clearly shown by the use of the enzyme α-mannosidase. This enzyme cleaved the sugar into mannose and cellobiose demonstrating that it is indeed α-mannosyl cellobiose of the structure shown. (Figure 5)

Using a similar approach the structures of the other mannose containing oligosaccharides were elucidated. In the case of the mannose containing disaccharide (E) the position of the mannosyl substituent was determined using the lead tetraacetate oxidation method described by Perlin. (8) On hydrolysis of the oxidised disaccharide, arabinose was detected, showing that mannose was linked to O_3. (Figure 6) The branched trisaccharide (B) was unexpectedly resistant to the action of both α-mannosidase and β-glucosidase presumably through steric hinderance of adjacent hexoses on O_3 and O_4 and therefore partial acid hydrolysis was used for this facet of the structural investigation. The acidic disaccharide (A) was shown to contain glucuronic acid and mannose in roughly equal proportions and was assumed to be the aldobiuronic acid previously isolated from the gum. Oligosaccharide D was shown to be cellobiose by co-chromatography with an authentic sample on paper and gas chromatography. (Figure 7)

All of the sugars in the newly proposed repeating unit of the polysaccharide with the single exception of the terminal mannose residue are represented in at least one of the oligomers, and our results are entirely consistent with the revised structure. (Figure 8)

It is possible that the covalent structures of gums produced under different conditions may vary in some way, for example, after chemical treatment. (10) There is evidence also that structural variation may occur in gums from different species of xanthomonas (9). Variation in structure is likely to be associated with variation in physical properties and it is possible that a range of xanthan gum types could be developed to give a

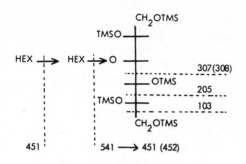

Figure 3. *Mass spectroscopy of the per-O-trimethylsilyl ether of the derived glycitol from oligosaccharide C (Figures in parentheses are after NaBD$_4$ reduction)*

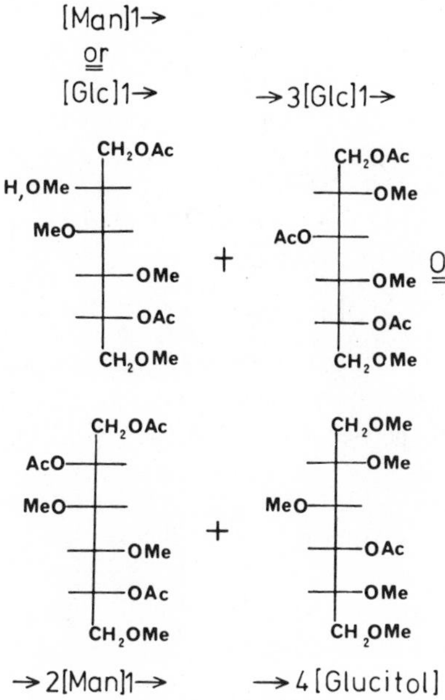

[Man]1→

or

[Glc]1→   →3[Glc]1→

Figure 4. *Partially methylated alditol acetates possible from gas chromatographic evidence*

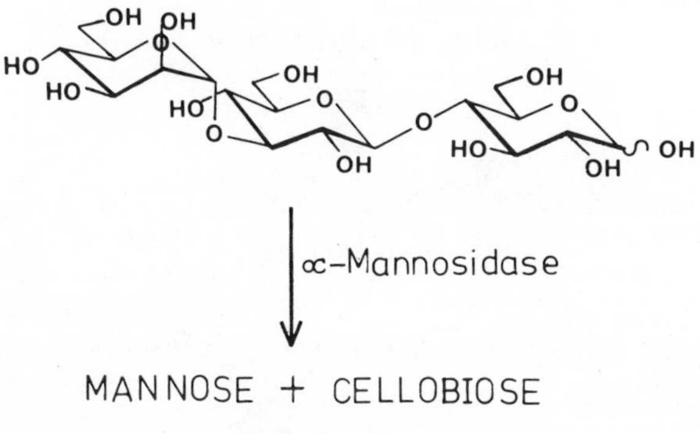

→2[Man]1→   →4[Glucitol]

α-Mannosidase

MANNOSE + CELLOBIOSE

Figure 5. *Action of α-mannosidase on oligosaccharide C*

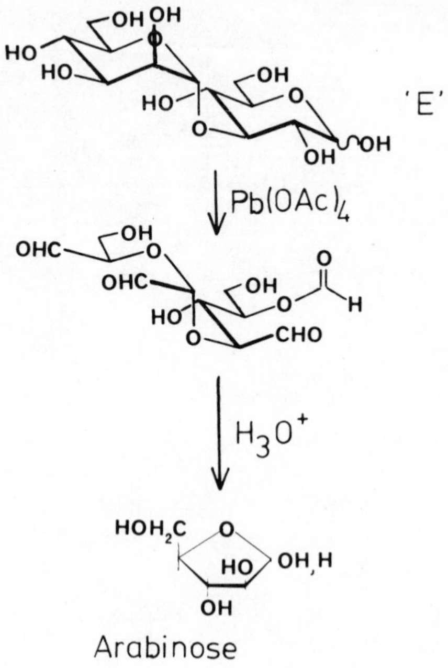

Figure 6. Lead tetraacetate oxidation of disaccharide E

Arabinose

A.   β-D-GlcAp-(1→2)-D-Manp

B.   B-D-Glcp-(1→4)-D-Glcp
                    |
                    ~
                    3
                    ↑
                    1
                    ⌣
                    |
              α-D-Manp

C.   B-D-Glcp-(1→4)-D-Glcp
                    |
                    ~
                    3
                    ↑
                    1
                    ⌣
                    |
              α-D-Manp

D.   β-D-Glcp-(1→4)-D-Glcp

Figure 7. Oligosaccharides from acetolysis of xanthan gum

E.   α-D-Manp-(1→3)-D-Glcp

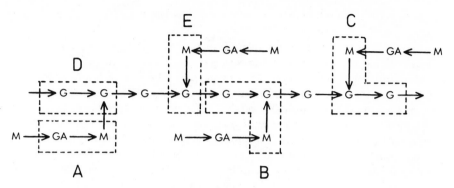

Figure 8. Location of acetolysis oligosaccharides in the repeating unit of xanthan gum

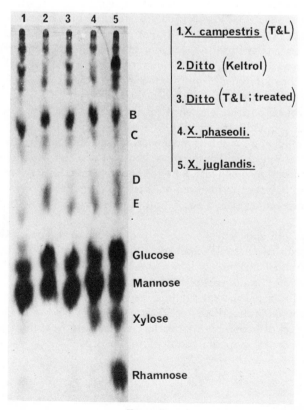

1. X. campestris (T&L)

2. Ditto (Keltrol)

3. Ditto (T&L ; treated)

4. X. phaseoli.

5. X. juglandis.

B
C

D

E

Glucose

Mannose

Xylose

Rhamnose

Figure 9.

range of products analogous to other gums in the plant series, such as the alginates and carrageenans.

Having characterised the five fragments of xanthan gum as described, it was realised that a means of comparing at least some facets of the covalent structural features of different samples might be offered, which could be accomplished in a relatively short period of time. Five samples of gum were therefore obtained namely (1) Xanthan gum produced by Tate and Lyle (2) Keltrol (3) a sample of gum containing culture broth treated with alkali at pH 12.2 at 83°C as described by Patton (4) Xanthomonas phaseoli polysaccharide obtained from Dr. P. Sandford NRRL Peoria and (5) A gum from Xanthomonas juglandis obtained from Dr. D. Ellwood of MRE Porton. Small samples of the above gums were treated as already described and the acetolysates worked up simultaneously through deacetylation and separation of neutral from acidic fractions. The neutral oligosaccharides were then examined on paper chromatography. (Figure 9)

All of the mannose containing oligosaccharides were revealed in similar proportion in each polysaccharide. Cellobiose is shown incompletely resolved and it appears to vary slightly in its proportion. A slow moving sequence of spots can be seen which are probably higher neutral oligosaccharides or possibly charged oligomers which have come through the resin treatment. The latter appear in approximately the same proportion in each polysaccharide. A faster moving spot thought to be xylose can be seen in the chromatogram of the Xanthomonas phaseoli polysaccharide. This latter sugar is possibly associated with the xanthan-like polysaccharide but there is as yet no positive evidence for this. The sample of gum from the Xanthomonas juglandis has not only the addition of xylose, but also rhamnose. The evidence for this is much more conclusive than the evidence for the identity of xylose in Xanthomonas phaseoli and the two sugars can be removed by fractionation with cetavlon showing conclusively that the xanthan-like polysaccharide does not contain rhamnose or xylose. The sugars do however appear to be present as polysaccharides as they are non-dialysable.

Finally, although not shown, the proportions of the acid oligosaccharides appeared similar in all samples and dominated by the aldobiuronic acid, as shown by paper chromatography of the unseparated acetolysates in oligomers located using neutral lead acetate spray.

In conclusion, while all findings are necessarily qualitative, the acetolysis approach described, provides a method for comparing samples of xanthan gum although no information is available from the terminal mannose unit or the pyruvate and acetate substituents.

Literature Cited

1. Sloneker, J.H. & Jeanes, A. Canad J Chem (1962) 40 2066
2. Siddiqui, I.R. Carbohyd Res (1967) 4 284
3. Imrie, F.K.E. British Patent 1331771
4. Morgan, K. & O'Neill, A.N. Canad J Chem (1959) 37 1201
5. Blake, J.D. & Richards, N.G. Carbohyd Res (1970) 14 375
6. Melton, L.D. (The Late), Mindt, L., Rees, D.A. & Sanderson, G.R. Carbohyd Res (1976) 46 245
7. Jansson, P.E., Kenne, L. & Lindberg, B. Carbohyd Res (1975) 45 275
8. Perlin, A.S. Anal Chem (1965) 27 396
9. Evans, C.E. Unpublished Results
10. Patton, J.T. (1970) United States Patent 3, 729, 460

15

Variation in *Xanthomonas campestris* NRRL B-1459: Characterization of Xanthan Products of Differing Pyruvic Acid Content

P. A. SANDFORD, J. E. PITTSLEY, C. A. KNUTSON,
P. R. WATSON, M. C. CADMUS and A. JEANES

Northern Regional Research Center, Agricultural Research Service,
U.S. Department of Agriculture, Peoria, IL 61604

Xanthan, the exocellular anionic heteropolysaccharide from Xanthomonas campestris NRRL B-1459 now produced industrially in both the United States (1) and Europe (2), has numerous applications in food and nonfood industries (3, 4). Xanthan is composed of D-glucose (Glc), D-mannose (Man), and D-glucuronic acid (GlcA) in the ratio of 2:2:1 (5-7) and of varying amounts of pyruvic and acetic acid (4, 5). Early structural analyses (8, 9) and more recent studies (10-13) indicate that xanthan consists of repeating pentasaccharide units (Figure 1). Upon a cellulosic backbone, trisaccharide side chains composed of β-D-Man(1→4)-β-D-GlcA(1→2)-α-D-Man are glycosidically linked to alternate glucose units at the 3-O-position. Acetic acid is attached as an ester to the 6-O-position of the internal mannose of the side chain (10) and pyruvic acid is condensed as a ketal with terminal mannose units (10, 14, 15). Recently, various substrains have been found (16-18) in certain stock cultures of the bacterium Xanthomonas campestris NRRL B-1459 that produce xanthan differing in yield, viscosity, various solution properties, and in pyruvic acid and acetyl content. These preliminary studies suggested that differences in pyruvic acid content were the main cause of these observed variations. Therefore, we re-examined the solution properties of two xanthan samples of differing pyruvic acid content at lower polysaccharide concentration and also examined xanthan samples of intermediate pyruvic acid content. Most xanthan applications are based on its unusual rheological properties (5, 19, 20); therefore, the differing rheological behavior of xanthans of differing pyruvic acid content has practical significance.

Experimental

Materials

The standard reference samples are those described previously (16) as PS-L and PS-Sm but they now are given the designation HPXan (for high pyruvate xanthan) and LPXan (for low pyruvate xanthan), respectively.

Laboratory Purification of Xanthan

The production and recovery of xanthan from broth were as previously reported (16). Xanthan from commercial sources was purified in a similar fashion but starting with a 0.25% dispersion. The yield of purified xanthan potassium salt from commercial sources was 70-85%.

Viscosity Measurements

Calibration of Viscometers. Standard oils of known viscosity were used to calibrate viscometers.

Viscosity Measurement at Xanthan Levels Above 0.25%. Viscosity measurements were made with a cone-plate micro visco-meter (Wells-Brookfield, Model RVT, 4.7 mm diameter and 1.565° angle cone) at 25° C and 1 rpm unless otherwise indicated. Dispersions for viscosity-concentration curves were prepared by volumetric, serial dilution, although the same results were obtained from individually prepared dispersions. Salt effects were observed by incremental addition of small amounts of solid salt to homogeneous, completely dispersed solutions of the polysaccharide. Readings usually were made after three revolutions, or when the values had become constant.

Viscosity Measurement at Xanthan Levels At or Below 0.1%. A Brookfield viscometer (model LVT) fitted with an Ultra-low (UL) adapter (Couette-type stainless-steel cell) was used to measure the viscosity of dilute solutions. Viscosity values (therefore shear rates) at 3.0 rpm with the UL adapter were closest to those obtained with either the cone-plate viscometer at 1 rpm or the LVT spindle (No. 3) at 30 rpm.

Viscosity vs Temperature. In the polysaccharide range of 0.25 to 2%, a Brookfield viscometer (model LVT) fitted with a No. 4 spindle was used to measure viscosities (30 rpm). Disper-sions in an 8-mm (inside diameter) tube, in which a thermocouple was placed in the dispersion to measure temperature, were heated in an oil bath over the temperature range of 2° to 95° C (above 95° C bubbles appear which lead to erratic readings). In the polysaccharide range of 0.1% or below, the UL adapter was placed

in an aluminum cylinder (1-cm walls) to distribute heat evenly.
Water was placed between UL adapter cup and the aluminum cylinder
to assure heat transfer. The aluminum cylinder was heated with
electrical resistance heating tape connected to a Variac. Temperature was measured with a thermocouple placed in the aluminum
cylinder.

Viscosity vs pH. The viscosity of 0.5% dispersions at
various pH's was measured with the cone-plate viscometer (1.0 rpm,
25° C) as previously described (16).

Intrinsic Viscosity

Size 75 Cannon semimicro viscometers (Cannon Instrument Co.)
were used to measure relative viscosity (n_{rel}) at 25° C.
Intrinsic viscosities, expressed as deciliters per gram (dl/g),
were determined by extrapolation of plots of $\frac{n_{rel}}{C}$ vs C to zero
concentration (C, g/100 ml).

Analytical Measurements

The method of Duckworth and Yaphe (21) was used for pyruvate
determination. Xanthan (3-5 mg) was hydrolyzed 3 hr at 100° in
2 ml 1 N HCl, neutralized with 2 ml 2 N Na_2CO_3, and diluted
to 10 ml with water. A 2-ml aliquot was pipetted into a quartz
cuvette with a 1-cm light path, and 1 ml of 1 N aqueous
triethanolamine buffer and 50-µl NADH solution (10 mg per ml of
1% $NaHCO_3$) were added. Absorbance (A) was measured at 340 nm and
4 µl lactate dehydrogenase (4,000 units per ml) were added.
Absorbance was measured again after 5 min, and at 5-min intervals
until stable. Percent pyruvate was calculated by the equation:

$$\% \text{ Pyruvate} = \frac{5 \text{ X } 88 \text{ X } 100 \text{ (A initial-A final) X } 3.05}{\text{Sample wt X } 6.22 \text{ X } 1,000}$$

where 88 is the molecular weight of pyruvic acid, 3.05 is solution
volume, 6.22 is the extinction coefficient of NADH, and 5 is a
dilution factor.

Q-Acetyl was determined by the hydroxamic acid method (22).

Component sugars in xanthan were determined by radiochromato-
graphic analysis of an acid hydrolysate after reduction with
^{3}H-sodium borohydride (23). D-Mannose and D-glucose content of
xanthan was independently checked by gas chromatography of their
alditol acetates (24). D-Glucuronic acid was also assayed by the
carbazole method (25).

Neutral equivalent weights were determined by titrating [with
standardized KOH (0.1 M)] decationized solutions (0.01 to 0.1%) (5).

Results

General Properties

The precipitation and rehydration behavior of xanthan products differ characteristically with pyruvic acid content of the
product. During precipitation with ethanol (2 volumes, also KCl,
1%), xanthan high in pyruvate (>4%) comes out of solution as a
cohesive stringy precipitate that tends to wind around the stirrer.
Under identical conditions, xanthan low in pyruvate (2.5 to 3.5%)
usually precipitates as a less cohesive particulate material that
does not wind on the stirrer. Brief heating of dispersions of
xanthan low in pyruvate (e.g., 0.1 to 1%, 3 min at 95° C) causes
precipitation behavior with alcohol and KCl to change to that like
xanthan high in pyruvate (see later). As isolated in the K-salt
form, both pyruvate types when freeze dried are white fibrous
products. Freeze-dried (K-salt form) HPXan products character--
istically take longer to rehydrate than low-pyruvate samples.
Apparently it is more difficult for water to completely penetrate
into HPXan. Dispersions of LPXan are generally clearer than HPXan,
which tend to have some opalescence, but this difference may
relate to the removal of cells and debris in our isolation
procedure.

Analytical Measurements

In Table I, the analytical results of HPXan and LPXan are
listed and compared to that expected for various theoretical xanthan structures of differing pyruvate content. When HPXan is
compared to LPXan only the amount of pyruvic and O-acetyl appear
to differ significantly, and perhaps the difference in O-acetyl is
not significant. The amount of D-glucose, D-mannose, and
D-glucuronic acid in both HPXan and LPXan are nearly identical.
The major compositional difference between these two types of
xanthan is in pyruvate content. The neutral equivalent weights of
HPXan and LPXan are consistent with their D-glucuronic and pyruvic
acid content. HPXan compares favorably to the theoretical repeat
unit depicted in Figure 1, in which there is an average of one
pyruvic acid ketal on every other terminal D-mannose unit; LPXan
is more like the theoretical repeat unit that has one pyruvate on
every fourth terminal D-mannose in the side chain.

Viscosity Measurements

Viscosity vs Polysaccharide Concentration. When compared at
polysaccharide concentrations of 1% or above, the viscosity of
LPXan is equal to or slightly higher than HPXan (see Figure 2).
At polysaccharide levels at and below 0.5%, LPXan is generally
less viscous than HPXan (Figure 2). Thus, the viscosity/
concentration curves of HPXan and LPXan cross near the 0.5%

EXTRACELLULAR MICROBIAL POLYSACCHARIDES

TABLE I   Comparison of pyruvic acid, acetyl, monosaccharide content, and neutral equivalent weight values of HPXan and LPXan to that of theoretical repeat units of differing pyruvic acid content.

Sample	Type	Pyruvic Acid g/100 g	O-Acetyl g/100 g	D-Glucose g/100 g	D-Mannose g/100 g	D-Glucuronic Acid g/100 g	Neutral Equivalent Weight
A. Experimental							
HPXan	High	4.4	4.5	37.0	43.4	19.5	633
LPXan	Low	2.5	3.7	37.7	42.9	19.3	790
B. Theoretical[1]							
Theory-1	Max[2]	8.7	4.3	35.6	35.6	19.2	506
Theory-2	High[3]	4.6	4.5	37.6	37.6	20.3	639
Theory-3	Low[4]	2.4	4.6	38.7	38.7	20.8	745
Theory-4	Min[5]	0	4.8	39.8	39.8	21.5	904

1/ Assume monosaccharide repeat unit as in Figure 1.
2/ Pyruvic acid ketal on every terminal D-mannose in side chain.
3/ Pyruvic acid ketal on every other terminal D-mannose in side chain.
4/ Pyruvic acid ketal on every fourth terminal D-mannose in side chain.
5/ No pyruvic acid ketal on terminal D-mannose in side chain.

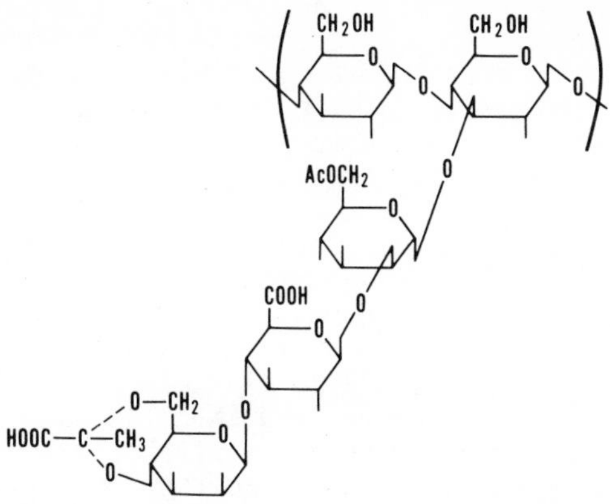

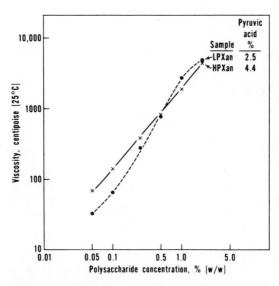

Figure 1. Structure of extracellular polysaccharide of Xanthomonas campestris according to Jansson et al. (10). Linkages denoted by – – – indicates pyruvic acid is not linked to every terminal D-mannose.

Figure 2. Viscosity vs. polysaccharide concentration. Viscosity of aqueous dispersions of the potassium salt-form of high-pyruvate xanthan (HPXan) (4.4% pyruvate) and low-pyruvate xanthan (LPXan) (2.5% pyruvate) were measured at 1 rpm (3.84 sec⁻¹) and 25°C.

polysaccharide level. At polysaccharide concentrations above
0.25%, both pyruvate types can be partially removed from solution
as a gel by centrifugation (100,000 X g, 60 min). Under similar
conditions but at lower concentrations (0.1%) polysaccharide of
either pyruvate type is not removed by centrifugation.

Viscosity vs Shear Rate. Both HPXan and LPXan display
recoverable shear-rate thinning at the polysaccharide concentrations tested. As shown in Figure 3, the viscosity decreases with
increasing shear rate. At the 1% level, HPXan shows slight
thixotropic behavior; i.e., previously sheared xanthan gives lower
viscosity values which recover with time. LPXan consistently
shows slight antithixotropic behavior at the 1% level. At lower
concentrations of both types, no thixotropy or antithixotropy is
observed.

Viscosity vs Temperature. Figure 4 illustrates the typical
viscosity behavior of HPXan and LPXan dispersions (1%) when
measured at various temperatures. Vastly different results are
found with the presence of KCl. When no added salt is present,
the viscosity of both pyruvate types starts to drop with increasing temperature. When the temperatures of the dispersions reach
about 50° C both types display viscosity changes in the opposite
direction (see curves C and D in Figure 4). With HPXan, a
dramatic rise in viscosity is seen, while with LPXan a slight
decrease is seen. With continued heating of the samples, HPXan's
viscosity reaches a maximum around 70° after which the viscosity
decreases rapidly. If the viscosity of the heated samples is
rechecked on cooling, the same viscosity peak is observed at the
temperatures 50-70°; in fact, this effect can be repeated over and
over with alternating heating and cooling. In rechecking the
viscosity of LPXan while it is cooling (see curve D, Figure 4), it
too now displays a sharp viscosity peak in the 50-80° temperature
range that was not seen in the initial heating curve. On
reheating, LPXan displays the viscosity peak in the 50-80° temperature range, as is observed with HPXan. The LPXan must be
heated above ∿60° before it displays the viscosity peak seen at
temperatures between 70-80° C. However, extended heating at 60°
for 8 hr did not produce the viscosity peak. Heating LPXan
dispersions at 95° for 3 min does produce the viscosity peak at
temperatures of 70-80°.

Heating LPXan alters its reactivity with salt. Figure 4
shows the effect of added salt on the viscosity at various temperatures. When 1% KCl is present, the viscosity of HPXan (curve A,
Figure 4) is nearly constant over the entire temperature range
between 10-90° C. The viscosity of 1% LPXan in the presence of
1% KCl decreases steadily with increasing temperature (curve B,
Figure 4). Only at 20-25° C are the viscosities of HPXan and
LPXan with 1% KCl similar. For high temperatures, e.g., 90° C,

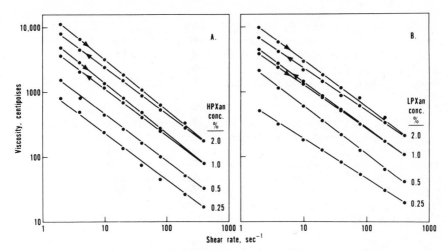

Figure 3. Viscosity (25°C) of xanthan dispersions vs. shear rate in sec⁻¹ at various poly-saccharide concentrations (w/v). (A) HPXan (4.4% pyruvate); (B) LPXan (2.5% pyruvate).

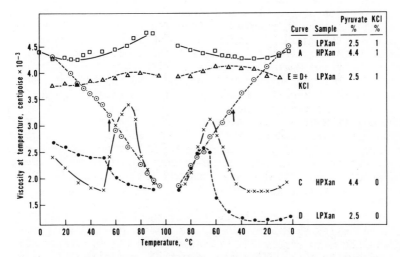

Figure 4. Viscosity of 1% dispersions of HPXan (4.4% pyruvate) and LPXan (2.5% pyruvate) at various temperatures with and without added KCl present. On left side of figure, the samples were gradually heated until they reached boiling. Then samples were allowed to cool and viscosity at temperature were remeasured (see right-hand side of figure). Spindle viscometer used.

the viscosity of LPXan is about 1/2 that of HPXan. If an LPXan
dispersion is first heated, such as with curve D, and then KCl (1%)
is added, the viscosity/temperature curve (E) that results is very
much like that for HPXan (curve A). However, if KCl (1%) is
present during the heating, no change to HPXan behavior is seen
(curve B).

At the 0.5% polysaccharide level (see Figure 5), the
viscosity-at-temperature behavior of both types is similar to that
obtained at the 1% level (Figure 4), but the viscosity peaks
appear at a lower temperature range (50-60°) and are much smaller.
As at the 1% polysaccharide level, heating salt-free dispersions
at the 0.5% level causes LPXan's behavior to become more like
HPXan; i.e., a viscosity spike at ∿50° appears after heating over
60° and its viscosity is greatly increased after heating when KCl
(1%) is added.

At the 0.1% polysaccharide level (Figure 6), the viscosity of
both pyruvate types decreases steadily with increasing temperature.
Heating of the LPXan solutions (no added KCl present) causes
subsequent cooling and reheating curves to be higher than initial
viscosity/temperature curve. Hence, heating changes LPXan to
resemble HPXan in behavior.

Effect of Salt on Viscosity. The viscosity of HPXan and
LPXan dispersions differs greatly in the presence of salt; however,
this difference is greatly diminished when salt-free LPXan dispersions are heated. In Figure 7, the effect of added KCl on the
viscosity of 1% and 0.5% dispersions of both pyruvate types are
compared with and without heating (95°, 3 min). When unheated
dispersions are compared, LPXan is less viscous, particularly at
high (1-3%) KCl concentrations. At the 0.5% polysaccharide level,
the viscosity of unheated LPXan is nearly unaffected by the
addition of KCl, whereas the viscosity of HPXan is nearly doubled
by the addition of as little as 0.4% KCl. Heating (95°, 3 min) of
a LPXan dispersion causes its behavior towards KCl to change to
that almost identical to HPXan provided KCl is not present during
heating but is present during the viscosity measurement. Heating
of HPXan under identical conditions has little effect on its
viscosity behavior towards added salt.

At the 0.1% polysaccharide level, the effect of heat and salt
on the viscosity of HPXan, LPXan, and a mixture of both pyruvate
types is shown in Figure 8. Unheated LPXan has significantly
lower viscosity than HPXAN (see Figure 8), sometimes 1/3 as much.
When heated (95°, 3 min), only LPXan's viscosity is affected.
Heating LPXan causes its viscosity to increase to or above that of
a HPXan whose viscosity is not affected appreciably (see Figure 8).
Also in Figure 8, the KCl/viscosity curves of a 1:1 mixture of the
HPXan and LPXan product are shown. The viscosity at 1% KCl of the

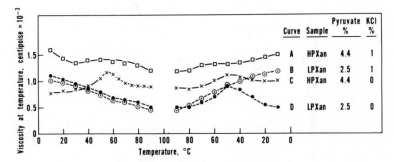

Figure 5. Viscosity of 0.5% dispersions of xanthan at various temperatures. See Figure 4 for experimental details.

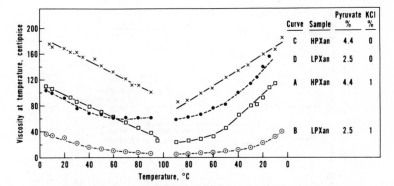

Figure 6. Viscosity of 0.1% solutions of HPXan (4.4% pyruvate) and LPXan (2.5% pyruvate) at various temperatures with and without added KCl present. Ultra-low-adapter of Brookfield LVT viscometer was heated in an aluminum cylinder; shear rate, 3.0 rpm.

202

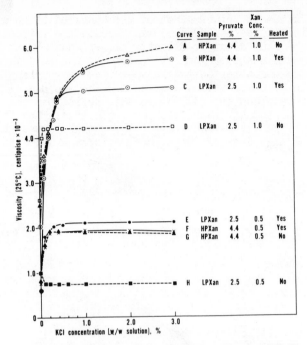

Figure 7. Effect of added salt (KCl) on viscosity (25°C,
3.84 sec⁻¹) of heated (95°C, 3 min) and unheated dispersions (1 and 0.5%) of HPXan (4.4% pyruvate) and
LPXan (2.5% pyruvate)

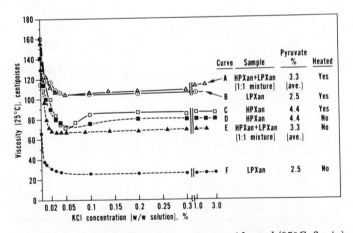

Figure 8. Effect of added salt on viscosity of heated (95°C, 3 min)
and unheated solutions (0.1%) of HPXan (4.4% pyruvate) and
LPXan (2.5% pyruvate). A 1:1 solution mixture of both pyruvate
types were tested also.

unheated mixture is between that of the unheated LPXan and HPXan of the mixture. Heating (95°, 3 min) causes the viscosity of the mixture to be equivalent or slightly higher than HPXan (heated or not heated) and heated LPXan. In Figure 9, the KCl/viscosity curves before and after heating dispersions of xanthans with intermediate pyruvate values are shown. The product with the highest pyruvate content (3.17%) in this series has the highest viscosity and products with lower pyruvate levels have correspondingly lower viscosities. Heating (95°, 3 min) causes the viscosity of all three of these intermediate pyruvate xanthans to increase.

Effect of pH on Viscosity. In Figure 10, the viscosities of 0.5% dispersions of HPXan and LPXan are compared as a function of pH. The viscosity obtained depends on the pyruvate type, the presence of additional salt, and for the LPXan whether or not the dispersion was heated. Typical U-shaped viscosity/pH curves obtained for HPXan (heated or not) and LPXan (heated) are flattened out by the addition of 1% KCl. Unheated LPXan dispersions gave inverted U-shaped viscosity pH curves that were not affected by the addition of KCl. Heating (95° C, 3 min) LPXan dispersions caused their viscosity/pH behavior to become more like HPXan.

Intrinsic Viscosity. When measured in water, the intrinsic viscosity was 102 dl/g for HPXan and 70 dl/g for LPXan (see Table II). When measured in ammonium acetate (0.01 M), a solvent previously found suitable for molecular weight studies (26), the intrinsic viscosity of HPXan was 43 dl/g while LPXan was 29 dl/g (see Table II). After heating dispersions of both pyruvate types separately (1%, 95°, 3 min) and diluting to proper concentration, the intrinsic viscosity value of HPXan was nearly as before heating, 42 dcl/g, but that for the LPXan increased to 39 dcl/g.

TABLE II   Intrinsic Viscosity of HPXan and LPXan.

No.	Pyruvic Acid Type	Pyruvic Acid g/100 g[3]	Intrinsic Viscosity[1] Not Heated Solvent Water	Intrinsic Viscosity[1] Not Heated Solvent 0.01 M NH_4Ac[4]	Heated[2] 0.01 M NH_4Ac[3]
1.	HPXan	4.4	102	43	42
2.	LPXan	2.5	70	29	39

[1] dl/g = deciliter per gram.
[2] Heated = 95° C, 3 min, 1% solution in water.
[3] Grams released by hydrolysis per 100 grams xanthan.
[4] NH_4Ac = ammonium acetate.

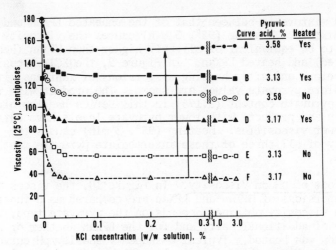

Figure 9. Xanthan products with intermediate (3.13% to 3.58%) levels of pyruvate. Viscosity vs. amount of added KCl of heated (95°C, 3 min) and unheated solutions.

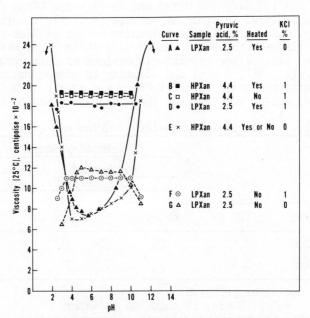

Figure 10. Viscosity (3.84 sec^{-1}, 25°C) vs. pH of dispersions (0.5%) of HPXan (4.4% pyruvate) and LPXan (2.5% pyruvate)

Birefringence. When viewed between crossed polarizers, dispersions of xanthan display birefringence, i.e., double refraction of light, even when an external orienting force is absent or very weak. Several factors such as rate of shear, concentration of polysaccharide, presence of extraneous salt, heat, and pH have previously been shown to affect this birefringence (27). When HPXan and LPXan pyruvate xanthans are compared (see Figures 11 and 12), the birefringence [retardation, (Δ) in nm] of unheated LPXan is much lower than that for HPXan, particularly at the low rpm's (see Figure 11). After heating (95°, 3 min) and cooling to 25° C, LPXan birefringence is increased and its birefringent behavior becomes much like that for HPXan. LPXan (1%) was heated to various temperatures, cooled to 25° C, and its retardation at zero rpm was measured (see Figure 12). This study shows that temperatures above 60° C must be reached in order for the LPXan dispersion to display temperature-increased birefringence.

Discussion

The pyruvic acid content of xanthan is an indicator of its solution properties. All xanthans high in pyruvate (>4.0%) show similar solution properties which are significantly different from those of xanthans low in pyruvate (2.5 to 3.0). Plots (see Figures 13 and 14) of viscosity vs pyruvic acid content of various xanthan products indicate that viscosity increases consistently with corresponding increases in pyruvate content. These data indicate that samples of xanthan with a pyruvate content higher than now normally found might be expected to have higher solution viscosities. If every terminal mannose carried a pyruvic acid ketal, the pyruvate content would be 8.69% (see Table 1) which is nearly double that now called a "high pyruvate" sample. Likewise, the data in Figures 13 and 14 show that xanthans with low-pyruvate content would be significantly less viscous than samples with higher pyruvate content.

The temperature and salt dependence of viscosity is concentration dependent. At high polysaccharide concentrations the rheology of HPXan and LPXan is similar, while at low concentrations they differ. The molecules of these two pyruvate types evidently interact differently. The anionic carboxyl of the pyruvate, like that of the uronate, influences charge distribution throughout the macromolecule. However, the distribution of pyruvate in the side chains is not known. Although a regular distribution is generally assumed, there is no evidence to confirm this notion. All the pyruvate could be clustered regionally in each molecule.

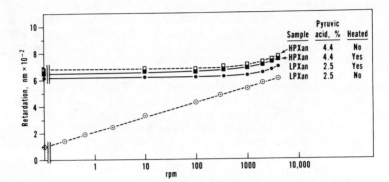

Figure 11. Retardation (Δ in nm) vs. rate of shear (rpm) of salt-free aqueous dispersions (1%) of HPXan (4.4% pyruvate) and LPXan (2.5% pyruvate)

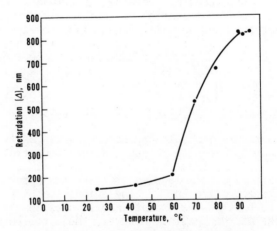

Figure 12. Retardation (Δ in nm) vs. temperature to which LPXan (2.51% pyruvate) dispersion (1%) was heated to before cooling to 25°C and measurement of Δ

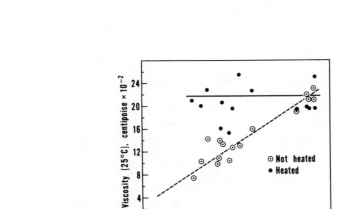

Figure 13. Viscosity (25°C, 3.84 sec⁻¹) vs. pyruvic acid content of xanthan. Dispersions, 0.5%, 1% KCl.

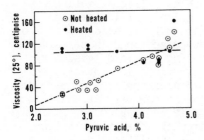

Figure 14. Viscosity (25°C, UL adapter, 3.0 rpm) vs. pyruvic acid content of xanthan. 0.1% solution, 1% KCl.

It should be noted that the pyruvate content of xanthans is segregated into two main groups, one at 2.5-3.5% pyruvate and another around 4.6% pyruvate. This grouping is perhaps significant in understanding the biosynthesis and source of pyruvate variability in xanthans produced by various substrains of Xanthomonas campestris B-1459.

Cause of the change in viscosity behavior upon heating aqueous dispersions of low pyruvate xanthan is not clear. The results could be interpreted as a new physical conformation being formed by the heating process. Alternatively, chemical changes could occur during the heating; e.g., introduction of cross linkages through ketal rearrangement, migration of acetic acid, or freeing of an esterified carboxyl group. However, the IR spectra of LPXan (and HPXan) that has been heated (95°, 3 min) is identical to spectra taken before heating. These studies also indicate that heating does not remove O-acetyl groups but they could migrate to other positions in the molecule.

Heating LPXan did cause its intrinsic viscosity value to increase to nearly that found for HPXan, whose value was not affected by heating. These data suggest that heating causes the molecular size, shape, or water-binding capacity of low pyruvate xanthan to become more like that found for HPXan.

Acknowledgment

We thank A. C. Eldridge for confirming by gas chromatography the neutral hexose content of several xanthan samples.

Abstract

Normal xanthan-producing strains of the bacterium Xanthomonas campestris NRRL B-1459 are characterized by their efficient conversion (~60%) of substrates such as D-glucose into extracellular polysaccharide that gives culture fluids of high viscosity (6,000 to 8,000 cpoise) and pyruvic acid contents of about 4.5%. Various substrates have been found in certain stock cultures that produce xanthan differing in yield, viscosity and other solution properties, and in pyruvic acid content. Analysis of xanthan products from these substrains and from commercial sources shows that the pyruvate content can vary at least from 2.5% to 4.8%, while the sugar composition (D-glucose, D-mannose, and D-glucuronic acid) remains constant. The precipitation, rehydration, and rheological behavior of all xanthan samples having high (4.0% to 4.8%) pyruvate were similar but significantly different from those samples having low (2.5% to 3.0%) pyruvate which display different properties. At xanthan concentrations of 0.1% to 0.5%, high pyruvate samples are more viscous (sometimes 2 to 3X more), particularly in the presence of salt, than low pyruvate samples. Brief heating (95° C, 3 min) of low-pyruvate solutions caused their solution properties to become more like high-pyruvate when observed in the presence of salt. Other rheological properties of both pyruvate types are examined.

Literature Cited

1. "Xanthan Gum/Keltrol/Kelzan/A Natural Biopolysaccharide for Scientific Water Control," Second Edition, pp. 1-36, Kelco Company, San Diego, California (1975).
2. Godet, P., Process Biochem., (1973) 8, 33.
3. Jeanes, A., J. Polym. Sci., Polym Symp., (1975), 45, 209.
4. Jeanes, A., Food Technol., (1974), 28, 34.
5. Jeanes, A., Pittsley, J. E., and Senti, F. R., J. Appl. Polym. Sci., (1961), 5, 519.
6. Sloneker, J. H., and Orentas, D. G., Nature, (1962), 194, 478.
7. Sloneker, J. H., and Jeanes, A., Can. J. Chem., (1962), 40, 2066.
8. Sloneker, J. H., Orentas, D. G., and Jeanes, A., Can. J. Chem., (1964), 42, 1261.
9. Siddiqui, I. R., Carbohydr. Res., (1967), 4, 284.
10. Jansson, P. -E., Kenne, L., and Lindberg, B., Carbohydr. Res., (1975), 45, 275.
11. Melton, L. D., Mindt, L., Rees, D. A., and Sanderson, G. R., Carbohydr. Res., (1976), 46, 245.
12. Lawson, C. J., and Symes, K. C., this symposium.
13 Moorhouse, R., Walkinshaw, M. D., Winter, W. T., and Arnott, S., Abstr. Papers Am. Chem. Soc. Meeting, (1976), 171, CELL 97.
14. Sloneker, J. H., and Orentas, D. G., Can. J. Chem., (1962), 40, 2188.
15. Gorin, P. A. J., Ishikawa, T., Spencer, J. F. T., and Sloneker, J. H., Can. J. Chem., (1967), 45, 2005.
16. Cadmus, M. C., Rogovin, S. P., Burton, K. A., Pittsley, J. E., Knutson, C. A., and Jeanes, A., Can. J. Microbiol., (1976), 22, 942.
17. Cadmus, M. C., Burton, K. A., Herman, A. I., and Rogovin, S. Abstr. Paper Am. Soc. Microbiol. Meeting (1971), 71, A47.
18. Kidby, D., Sandford, P., and Herman, A., Appl. Environ. Microbiol., in press.
19. Holzworth, G., Biochemistry, in press.
20. Jeanes, A., and Pittsley, J. E., J. Appl. Polym. Sci., (1973), 17, 1621.
21. Duckworth, M., and Yaphe, W., Chem. Ind., (1970), 747.
22. McComb, E. A., and McCready, R. M., Anal. Chem., (1957), 29, 819.
23. Knutson, C. A., Carbohydr. Res., (1975), 43, 225.
24. Sawardeker, J. S., and Sloneker, J. H., Anal. Chem., (1965), 37, 945.

25. Knutson, C. A., and Jeanes, A., Anal. Biochem., (1968), 24,
    482.
26. Dintzis, F. R., Babcock, G. E., and Tobin, R., Carbohydr.
    Res., (1970), 13, 257.
27. Pittsley, J. E., Sloneker, J. H., and Jeanes, A., Abstr.
    Papers Am. Chem. Soc. Meeting (1970), 160, CARB 21.

Zanflo—A Novel Bacterial Heteropolysaccharide

K. S. KANG, G. T. VEEDER, and D. D. RICHEY

Kelco, Division of Merck & Co., Inc., 8225 Aero Drive, San Diego, CA 92123

An exocellular bacterial polysaccharide called "xanthan gum" that was originally developed by the Peoria Lab of the USDA nearly two decades ago is at this time the only bacterial heteropolysaccharide that is being produced on a large commercial scale. The commercial success of xanthan gum is attributed to its many valuable and often unique properties for industrial applications and to an economic manufacturing process.

Among our novel polysaccharides that are produced by many bacterial species that we have isolated in our screening program, a product which is now trade named ZANFLO is especially outstanding in its fermentation efficiency and product properties.

ZANFLO is produced by a bacterium that was isolated from a soil sample taken at Tahiti. The organism is a gram-negative, non-sporeforming rod with a size range of 0.75-1.0 by 1-2μ. The dimensions change during the fermentation. At the beginning of the fermentation they are large rods which quickly change to a coccobacillus 0.75-1.0μ in diameter. It is heavily encapsulated. Some of its biochemical characteristics are shown in Table I. This organism produces a positive lactose reaction within 24 hours. It possesses nitrate reductase, CMCase, urease and lysine decarboxylase. It can utilize citrate as a sole carbon source and will grow in the presence of 8% NaCl. Its optimum growth temperature is 30-33°C. and growth will occur at 45° C. In litmus milk this organism produces an acid curd with peptonization and reduction of the litmus. At the present time we are evaluating these and other significant taxonomic tests to ascertain the identity of this microorganism.

This organism is quite specific about what carbon source it will use for optimum polysaccharide production. It produces an excessive amount of acid with glucose as a carbon source with only minimum polysaccharide synthesis. Even with pH control the conversion efficiency with glucose is still very low. Polysaccharide synthesis is better with sucrose or maltose or mixtures of these than with glucose, but the conversion efficiency is still poor. Improved polysaccharide synthesis is found with

lactose and hydrolyzed starch. We typically hydrolyze our starch
slurries with commercially available α-amylase preparations.
The preferred medium contains phosphate as a buffering
agent, ammonium nitrate and a soy protein product as nitrogen
sources, and magnesium sulfate in addition to the carbon source.
The data in Figure 1 shows the results of a typical
batch-type fermentation in a pilot plant fermentator. The
inoculum size is typically 5% with a 3% (as is) carbon source
concentration. Viscosity development started at approximately 7
hours and reached a maximum of 4500 cps by 64 hours. Unless
otherwise specified, viscosity measurements are made using a
Model LVF Brookfield viscometer with #4 spindle at 60 rpm. The
maximum cell population of about 1 x 10^{10} was reached in 10
hours.

By employing an automatic agitation and aeration controlling
system with an oxygen probe, the minimum dissolved oxygen concentration was determined to be 5-10% during the first 24 hours of
the fermentation.

Recovery of the product is done by precipitation with such
organic solvents as acetone, ethyl alcohol, isopropyl alcohol,
or various isomers of butanol. After precipitation, the polymer
fibers are dried and milled to a powder.

The studies on the chemical components of this polysaccharide were done on material purified by filtration of ZANFLO
solutions using diatomaceous earth as filter aid followed by
repeated reprecipitations with IPA. The material was found to
be 97% carbohydrate and 3% protein. The polysaccharide was
hydrolyzed using 2N H_2SO_4 and heated to 100° C. for 5 hours.
The components of this polymer were identified using paper
chromatographic techniques. The molar ratio was determined
using gas liquid chromatography. In Table II one can see the
results of this chemical component analysis. The carbohydrate
portion was found to contain glucose, galactose, glucuronic
acid, and fucose in the molar ratio of 3:2:1.5:1. Uronic acid
accounts for approximately 20% of the polysaccharide on a weight
basis. It is noteworthy that fucose is not commonly found as
a structural constituent of exocellular bacterial heteropoly-
saccharides. We are not yet certain as to the role of protein
in the purified polysaccharide. The completion of our structural
work will provide the answers to these questions.

ZANFLO is a high-viscosity polysaccharide, as shown in the
viscosity concentration curve in Figure 2. It is considerably
higher than that of xanthan gum, a well-known bacterial heteropolysaccharide widely used in industry. This difference becomes
more outstanding at higher concentrations. At a 1.5% gum concentration, the xanthan gum had a viscosity of 2500 cps, while
ZANFLO had a viscosity of 5000 cps.

The results in Figure 3 show the effect of heat on the
ZANFLO polysaccharide. ZANFLO's viscosity, like that of many
microbial polysaccharides, is definitely affected by heat. As

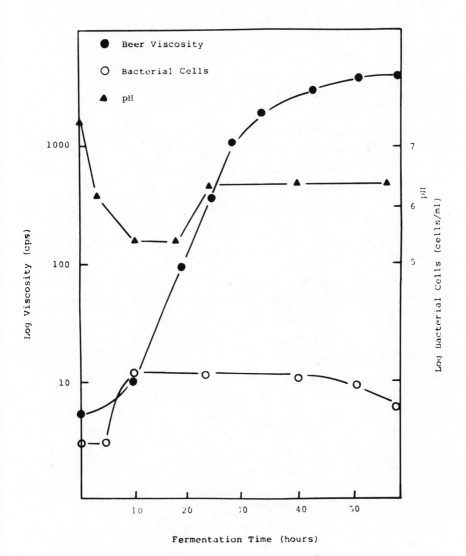

Figure 1. *Fermentation parameters of the Zanflo fermentation*

Table I. Biochemical Characteristics

NITRATE REDUCTASE	+
CELLULASE (Cx)	+
UREASE	+
AMYLASE	±
H₂S PRODUCTION	+
INDOLE	−
VOGES-PROSKAUER	+
METHYL RED	−
LYSINE DECARBOXYLASE	+
GELATIN LIQUEFACTION	−
ACID AND GAS FROM CARBOHYDRATES	
GLUCOSE	+
LACTOSE	+
SUCROSE	+
MALTOSE	+
CELLOBIOSE	+
MANNITOL	+
INOSITOL	+
ADONITOL	−
DULCITOL	−

Table II. Carbohydrate
Composition of Zanflo

URONIC ACID	19%
GLUCOSE	39%
GALACTOSE	29%
FUCOSE	13%
	100%
ACETYL	4.5%

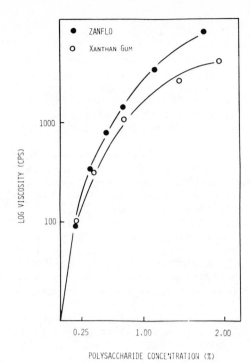

Figure 2. *Viscosity vs. concentration*

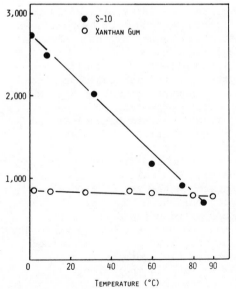

Figure 3. *Effect of temperature on Zanflo viscosity*

you can see, the relationship between viscosity and temperature
is quite linear. From this graph one can calculate a decrease
in viscosity of 25 cps/°C. as the temperature is increased. The
ZANFLO concentration is 1%. The viscosity decrease is temperature reversible.

The results in Figure 4 show the effect of pH on ZANFLO in
comparison to that of xanthan gum. The viscosity of ZANFLO
remains stable from a pH of 5 to 10 but decreases on either side
of this range.

One of the most striking properties of this polysaccharide
is its compatibility with cationic dyes. Anionic gums, such as
xanthan gum and many uronic acid-containing polysaccharides,
react strongly with cationic dyes, such as methylene blue
chloride, to form a fibrous precipitate which limits their
industrial applications. However, ZANFLO, even though it contains a substantial amount of uronic acid, does not precipitate
with such cationic dyes at any pH.

It was noted, nonetheless, that ZANFLO lost all or most of
its viscosity during the test and further experimentation showed
the reaction with methylene blue to be influenced by salt concentration and pH.

The results in Table III demonstrate the effect of pH on the
viscosity of the ZANFLO-methylene blue complex using acetic acid
to adjust the pH. The viscosity from neutrality to at least 4.5
is as low as water viscosity. By the time a pH of 3.9 is
reached, the viscosity starts to increase and reaches a maximum
of 110 cps at a pH of 3.1. We also noticed at this time that
the viscosity would decrease to that of water again if the pH
were adjusted upward slowly with NaOH.

If one continues to add NaOH, the viscosity will begin to
increase at a pH of 4.4-4.6 and it can no longer be brought down
to that of water again. This effect is caused by the concentration of Na^+ now present in the solution.

The results in Figure 5 show the effect of monovalent and
divalent cations on a similar system. Here, various concentrations of NaCl or $MgCl_2$ were added to neutral solutions of ZANFLO
containing MBC and having the viscosity of water. The results
show that a stoichiometric relationship exists between Mg and
Na in restoring lost viscosity to the solution. It appears
evident that an increase in electrolyte concentration would
increase competition with MB for the reactive site of polymer
and therefore increase the viscosity. The role of pH in this
phenomenon appears to be centered around the pKa of the uronic
acid portion of the polymer. Glucuronic acid has a pKa of 3.2.
At pH values below this, the number of reactive sites of the
polymer from MB would decrease, and, hence, increase in viscosity.

The mechanism involved in compatibility is not understood
at present. We speculate that perhaps the protein or peptide
moiety may play an important role in masking the anionic groups
of the gum or stabilizing the gum under these conditions.

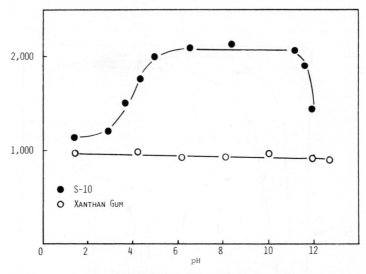

Figure 4. Effect of pH on Zanflo viscosity

**Table III. Effect of pH on the
Viscosity of a Zanflo-Methylene
Blue Chloride Complex
in Solution**

pH	VISCOSITY
7.5	440 CPS (NO MBC)
7.5	0 CPS (0.2% MBC)
5.8	0 CPS
4.2	0 CPS
3.9	25 CPS
3.1	110 CPS

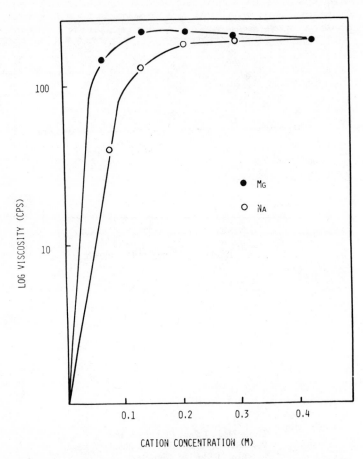

Figure 5. Effect of monovalent and divalent cations on restoring viscosity to Zanflo-MBC solutions

Another possibility is that the secondary or tertiary structure of this polysaccharide, which would give it a distinctive conformation in solution, protects it against the action of cationic agents such as methylene blue chloride.

In summary, we have isolated a bacterium from the soil which produces large amounts of a novel industrial heteropolysaccharide. Because of its unusual rheological properties and compatibility with basic dyes, ZANFLO has already established excellent applications in paint and shows excellent potential in other application areas.

17

PS-7—A New Bacterial Heteropolysaccharide

K. S. KANG and W. H. McNEELY

Kelco, Division of Merck & Co., Inc., 8225 Aero Drive, San Diego, CA 92123

Dextran production by fermentation or by cell-free synthesis has long been known (1). However, polysaccharide fermentation of other types is a relatively new field. The Peoria Laboratory of the USDA, which carried out the initial microbial research on xanthan gum, also has done development work on polysaccharide production by various Arthrobacter and yeast strains (2).

Dextran, xanthan gum, polytran, and ZANFLO represent microbial polysaccharides which are now commercially available. The most notable fermentation polysaccharide with the greatest commercial success at this time is xanthan gum.

This paper is intended to report on the various properties in relation to possible industrial applications of a novel polysaccharide, PS-7. The microbiological and fermentational aspects of this polysaccharide are described only briefly. A preliminary report of PS-7 as a potential food additive has been made (3).

PS-7 is produced by a soil bacterium which was isolated from a local soil sample. This organism was identified as a strain of Azotobacter indicus on the basis of growth characteristics, biochemical and morphological properties.

PS-7 is an extracellular polysaccharide produced by means of an aerobic, submerged fermentation. The typical fermentation medium is illustrated in Table I. The concentration of potassium phosphate may be reduced to as low as 0.01% if the pH has been maintained during the fermentation, ordinarily by using KOH.

By employing an automatic agitation-controlling system with an oxygen probe, the minimum dissolved oxygen concentration was determined to be 5-10% during the first 20-25 hours of the fermentation. During this fermentation time with constant aeration rate of 0.25

liter per liter of fermentation medium per minute, the
tip speed for agitation varied between 60 and 110
meters per minute to maintain the D.O. of 10%. The
viscosity of the fermentation liquor steadily increased, and, at 24 hours, the viscosity was approximately 2000 cps, as measured by a Brookfield viscometer
at a rotation speed of 60 rpm and ambient temperature.

PS-7 may be recovered from the fermentation liquor
by alcohols such as methanol, ethanol and isopropanol,
etc., or by lower alkyl ketones such as acetone. The
preferred solvent for recovery purposes is IPA. After
precipitation, the polymer fibers are dried and milled
to obtain a pale, cream-colored powder.

The chemical component analyses by paper chromatography and gas-liquid chromatography were carried
out and the results are shown in Table II. The results
indicate that PS-7 consists of glucose, rhamnose and
a uronic acid in an approximate ratio of 6.6:1.5:1.0.
It has an acetyl content of about 8.0-10.0%. Neither
the uronic acid nor the linkages present in this
polymer have been elucidated at this time.

PS-7 has an unusually high viscosity, as demonstrated in Figure 1. The viscosity of PS-7 is much
greater than that of xanthan gum, with the difference
becoming significant at concentrations as low as 0.15%.
This point is further illustrated in Figure 2. The
fermentation liquor in the flask becomes so viscous
at the end of the fermentation time that there is
little flow, even if the flask is held upside down.

Figure 3 illustrates another important property
of any commercial polysaccharide, and that is its
viscosity response to temperature. This data shows
that PS-7 is similar to xanthan gum in that its viscosity is stable over a wide temperature range.

The viscosity response of PS-7 to changes in pH
is shown in Figure 4. The viscosity of PS-7 is almost
as stable as xanthan gum, although the viscosity starts
to decrease below pH 3 and beyond pH 12.

Another important characteristic for some applications is pseudoplasticity. Pseudoplasticity is indicated
when the viscosity decreases as the shear rate is increased. This property of PS-7 is depicted in Figure 5.
This viscogram was obtained by using the Fann viscometer Model 35. It should be noted that the concentration of PS-7 is only half the concentration of xanthan
gum or a quarter of other polymers such as CMC, HEC, and
guar. There is a remarkable decrease in the viscosity
of PS-7 as it is sheared, the magnitude well exceeding
that of xanthan gum. Figure 6 compares viscosity and
pseudoplasticity of PS-7 to xanthan gum at the same

Table I

TYPICAL FERMENTATION MEDIUM

Ingredients	Concentration
K$_2$HPO$_4$	5.0 grams
MgSO$_4$ · 7 H$_2$SO$_4$	0.1 grams
NH$_4$NO$_3$	0.9 grams
PROMOSOY	0.5 grams
GLUCOSE	30.0 grams
TAP WATER	to 1 liter

Table II

THE CHEMICAL COMPONENTS OF PS-7

CARBOHYDRATE COMPOSITION

	PS-7
URONIC ACID	11%
GLUCOSE	73
RHAMNOSE	16
	100%
ACETYL	9%

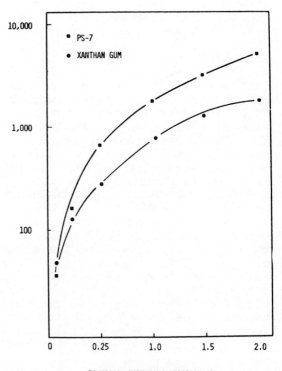

Figure 1. The viscosity of PS-7

POLYSACCHARIDE CONCENTRATION (%)

Figure 2

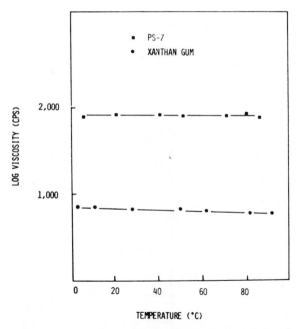

Figure 3.   The effect of heat on the viscosity of PS-7 and
xanthan gum solutions

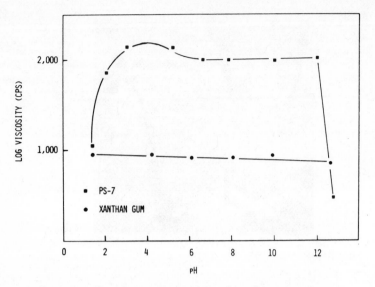

Figure 4.   The effect of pH on PS-7 and xanthan gum solutions

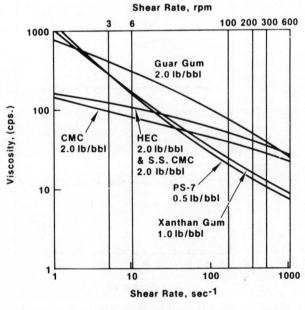

Figure 5

polymer concentration. Here, the differences in both
viscosity and pseudoplasticity between PS-7 and xanthan
gum are more outstanding than in Figure 5.

PS-7 is fully soluble hot or cold in distilled
water, tap water, brine water or sea water. This fact
is illustrated in Table III. The sea water has a salt
concentration of approximately 3.5%, and the Permian
Brine water of West Texas has a salt content of approximately 26%. The lowest viscosity was obtained in
distilled water.

Polysaccharide PS-7 is also compatible with a wide
variety of salts. Examples of this compatibility are
shown in Table IV. These concentrations are not
necessarily the limits for PS-7 compatibility. The
stability of these solutions was checked after three
hours stirring and 24 hours standing, looking for
precipitation, gelation or changes in flow properties.
All solutions listed in this slide were stable.

The stability of PS-7 was also studied over a
one-month period in many of these salts, and this data
is shown in Table V. A distilled water control was
carried. As a preservative, formaldehyde was added
to the solutions at the concentration of 200 ppm. The
data indicates that PS-7 was stable in all cases.

Polysaccharide PS-7 is incompatible with cationic
or polyvalent ions at high pH. This incompatiblity
results in a gel. Solutions of PS-7 also exhibit some
tendency to gel in the presence of high concentrations
of monovalent salts above a pH of 10. While this
gelation is considered an incompatibility, it can also
be a desirable effect, as will be shown later. Solutions of PS-7 show limited storage stability under
conditions of strong acidity or alkalinity.

The properties illustrated in the previous slides
demonstrate many characteristics of PS-7 which make it
a highly useful agent in oil well drilling muds.
Drilling muds are generally aqueous fluids which contain substantial quantities of clays and other colloidal materials. An optimum drilling fluid would be one
which, firstly, is flexible in its viscosity characteristics so as to provide suspension of solids within
the fluid, and, secondly, would lubricate the drill bit.
The high viscosity, pseudoplasticity and relative insensitivity of the viscosity to temperature indicate
that the use of PS-7 in a drilling mud would come close
to the optimum characteristics discussed earlier.

A very simple, easy and rapid way to test the
suspending ability of a fluid was developed in our
laboratory (4). The method employs a standard American
Petroleum Institute sand-content tube. The tube is

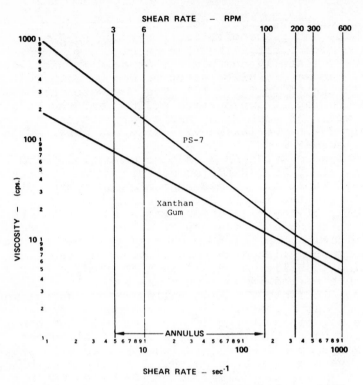

Figure 6. Viscogram of PS-7. Concentration of polymers, 0.14%.

Table III

VISCOSITY OF PS-7 IN VARIOUS WATERS
OF A 0.5% CONCENTRATION

WATER	VISCOSITY (CPS)
DISTILLED	690 CPS
TAP WATER	820 CPS
SEA WATER	860 CPS
PERMIAN BRINE	720 CPS

Table IV

Salt	Final Salt Concn.
Aluminum Nitrate	(40%)
Aluminum Sulfate	(25%)
Ammonium Chloride	(30%)
Ammonium Nitrate	(65%)
Ammonium Sulfate	(22%)
Calcium Chloride	(32%)
Calcium Nitrate	(55%)
Magnesium Chloride	(36%)
Magnesium Nitrate	(44%)
Magnesium Sulfate	(30%)
Potassium Ferricyanide	(7%)
Potassium Ferrocyanide	(25%)
Sodium Chloride	(26%)
Sodium Dichromate	(16%)
Sodium Nitrate	(40%)
Sodium Phosphate (dibasic)	(10%)
Sodium Sulfate	(15%)
Sodium Sulfite	(20%)
Sodium Thiosulfate	(40%)
Zinc Chloride	(40%)
Zinc Sulfate	(37%)

Table V

SOLUTION STABILITY OF PS-7

1% PS-7 SOLUTION WITH	VISCOSITY (CPS)	
	INITIAL	AGED 1 MO.
DISTILLED WATER (CONTROL)	1700	1900
SODIUM CHLORIDE (15%)	3150	3450
CALCIUM CHLORIDE (15%)	3050	3250
ALUMINUM SULFATE (15%)	3000	3100
ZINC SULFATE (15%)	3100	3200
AMMONIUM CHLORIDE (23%)	2850	3100
CUPRIC CHLORIDE (13%)	2950	2850
FERROUS SULFATE (13%)	3100	2650
MONOSODIUM PHOSPHATE (13%)	3050	3550
ZINC CHLORIDE (13%)	3000	3350

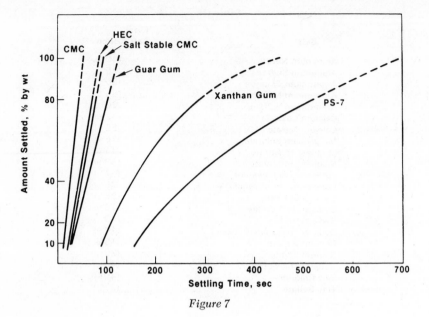

Figure 7

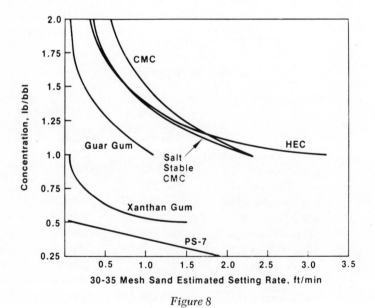

Figure 8

filled with the fluid to be tested. Then, 1 gram of
20-35 mesh sand is placed into the tube. The tube is
shaken vigorously to thoroughly disperse the sand,
placed upright, and timing is begun. The time that
it takes for the sand to reach selected graduated marks
at the bottom of the tube is recorded.

Figure 7 is a plot of the settling rates of
particulate matter in the different fluids. The data
indicate that PS-7 was by far the most efficient in
suspending particles. Guar gum was intermediate, and
the cellulosics the least effective.

Figure 8 is a plot of effective settling rate
vs. concentration for the fluids. Here, again, PS-7
was the most effective, and it takes approximately
half the concentration of xanthan gum to produce comparable suspending effect.

Drilling fluids are typically prepared with whatever water is available nearby. The good solubility
of PS-7 makes it suitable to this application--either
the sea water for off-shore wells or fresh or brine
water from water wells at the drilling site, whichever
may be available. In some drilling fluids, it is desirable to gel the fluid. Polysaccharide PS-7 is also
useful in this respect, as increasing the pH to 11
causes a drilling fluid containing PS-7 and chromium
to gel.

The viscosity of the drilling fluid can be adjusted to the desired level by simply altering the pH
and the concentration of the cross-linking agent.
Besides the oil well drilling applications, we have
established the potential use of PS-7 in other applications such as dripless water-base latex paint,
waterflooding systems for secondary oil recovery, wall
joint cement adhesives and textile printing.

Abstract

PS-7 is an anionic heteropolysaccharide produced by a
strain of Azotobacter indicus in an aerobic fermentation. PS-7 is composed of glucose, rhamnose, and
uronic acid in an approximate weight ratio of 6.6:1.5:1.
The polysaccharide has an acetyl content of about 9%.
Solutions of PS-7 are characterized by high viscosity
and a high degree of pseudoplasticity. The polysaccharide has excellent solubility in sea water and even
in brine containing 25% salt. PS-7 exhibits excellent
temperature and pH stability and is compatible with
a variety of salts. In the presence of Cr^{+++}, PS-7
gum can be cross-linked at pH 9.0-9.5. These properties indicate that PS-7 will find wide utility in a

variety of applications.

Literature Cited

(1)   Jeans, Allene, "Dextrans and Pullulans," Extra-
      cellular Microbial Polysaccharides of Practical
      Importance, ACS Symposium series (1976).

(2)   McNeely, William H. and Kang, Kenneth S., "Xan-
      than and Some Other Biosynthetic Gums," Industrial
      Gums (2nd Ed.), p. 473, Academic Press, New
      York (1973).

(3)   Kang, Kenneth S. and Kovacs, Peter, "New Micro-
      bial Polysaccharides as Potential Food Additives,"
      IVth International Congress of Food Sci. and
      Technol., Madrid, Spain (1974).

(4)   Carico, Robert D., "New Field Test Improves
      Fluid-Suspension Measurements," Oil and Gas
      Journal, (1976) 74, (27), p. 81.

Applications of Xanthan Gum in Foods and Related Products

THOMAS R. ANDREW

Kelco, Division of Merck & Co., Inc., 8225 Aero Drive, San Diego, CA 92123

Xanthan gum was approved for use as a food additive on March 19, 1969, by the Food and Drug Administration in accordance with 21 CFR 121.1224. Since then it has been used in a wide variety of foods for a number of important reasons including emulsion stabilization, temperature stability, compatibility with food ingredients, and its unique pseudoplastic rheological properties.

It has been proposed that xanthan gum, which is produced by the microorganism Xanthomonas campestris, is in fact a survival device for the organism generating it and it has, through millions of years of evolution, been perfected for this purpose (1). Thus, it is not surprising that a substance generated for the protection of a microorganism should possess such unusual properties and be so resistant to thermal, chemical, and biological degradation.

A great deal of progress has been made toward understanding the chemistry of xanthan gum. Figure 1 shows the structure of xanthan gum as initially postulated. It was shown that it consisted of a sixteen-residue repeating unit composed of D-glucose, D-mannose, and D-glucuronic acid as shown (2, 3). In a paper published in 1975 Jansson and co-workers proposed the somewhat more simplified structure shown in Figure 2. This structure is composed of a backbone of 1 - 4 linked β-D-glucose units with side chains which consist of two mannose and (4) a glucuronic acid unit on every other glucose unit. Every other side chain carries a pyruvic acid group.

In 1972 Rees proposed a double helix solution conformation (Figure 3) for xanthan gum which went a long way toward explaining the yield point phenomenon and the flat temperature-viscosity curve, unique among polysaccharides (5,6). A further extension of Rees's study (Figure 4) provided an

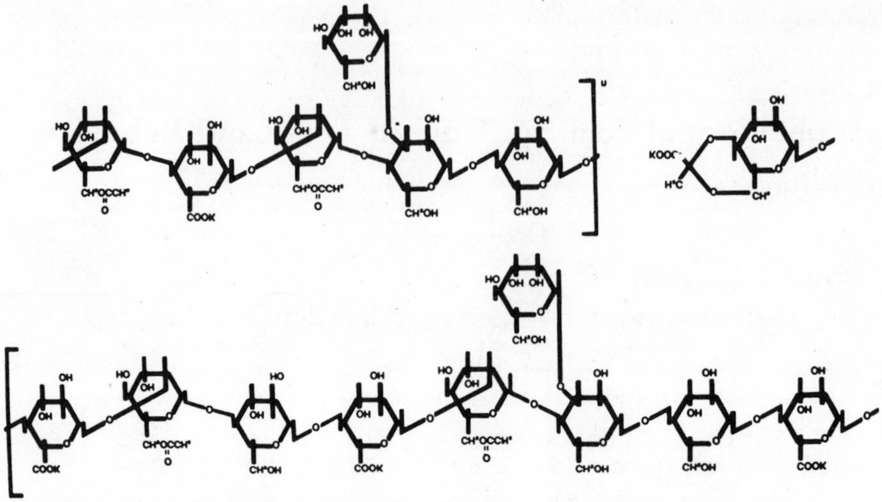

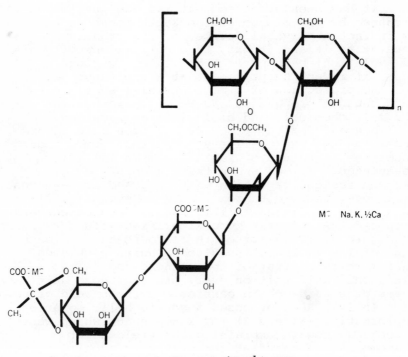

Figure 1. *Xanthan gum structure (pre 1975)*

Figure 2. *Structure of xanthan gum*

excellent explanation of why xanthan gum reacts
to form elastic gels with locust bean gum but not
guar gum. The uniformly distributed galactose side
chains along the mannose backbone of guar gum prevent
the close association of the molecule with the
xanthan gum helix while the existence of "smooth"
zones (Figure 5) in the locust bean gum molecule
allows association and, therefore, gelation (7).

     As is usually the case with the use of gums,
theory and practice have run on almost parallel
paths but the twain have not yet met. Theory helps
us understand the results but has not led us to them.

     When xanthan gum entered the food marketplace
in 1969, its basic physical properties were actively
promoted, followed by a number of suggested formula-
tions, all of which were examples of advances over
previous systems stabilized with other gums. Salad
dressing with emulsion stability extended to a year
or more and salad dressings that could be retorted
or repeatedly frozen were developed. An instant
pudding which was almost the same as the cooked starch
version was formulated, and industry developed a
number of other improved products.

Pasteurized Processed Cheese Spread

     More recent work has been completed which
demonstrates the utility of xanthan gum in pasteurized process cheese spread (8). Using a standard
formula for pasteurized process cheese, samples were
prepared under duplicate conditions in a cooker
commonly used for this purpose. As Table I
indicates, a number of combinations of xanthan gum,
locust bean gum, and guar gum were evaluated to
determine the optimum ratio for good meltdown,
firmness, sliceability, and flavor release. It is
interesting to note that only one combination gives
good results in every category, Trial No. 17A.
These results not only demonstrate the utility of
xanthan gum but also the necessity for using blends
at times to obtain functional advantages not
obtainable with single thickeners. Each gum contributes a desirable characteristic to the final
product, but the synergistic blend is better than
the sum of its parts.

Cottage Cheese Dressing

     Table II again illustrates the functional
superiority of the synergistic blend. Cottage cheese

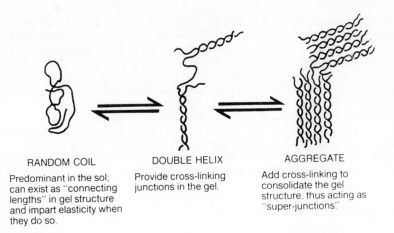

RANDOM COIL	DOUBLE HELIX	AGGREGATE
Predominant in the sol; can exist as "connecting lengths" in gel structure and impart elasticity when they do so.	Provide cross-linking junctions in the gel.	Add cross-linking to consolidate the gel structure, thus acting as "super-junctions."

Sol ⇌ Incipient gel ⇌ Clear elastic gel ⇌ Stiff gel ⇌ Turbid rigid gel ⇌ Phase separation syneresed gel

Figure 3. States of polysaccharide molecules and their role in gel properties

Figure 4. Schematic of galactomannan conformation. Each line represents a sugar unit consisting of the backbone composed of β-D-mannopyranose units and the side chains composed of α-D-galactopyranose units.

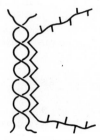

Figure 5. Possible model for the interaction between xanthan gum and locust bean galactomannan, resulting in gel formation

cream dressing, which is normally about 10 percent butterfat with salt added, is formulated to possess enough viscosity to make it creamier and to give good cling to the curd. If the viscosity of the cream is too high, the resulting cottage cheese will be too "dry" and the eating qualities will suffer. Another potential problem arises because of the inherent gummy mouthfeel exhibited by many gums at a concentration of 0.2 to 0.3 percent, the normal use level of gums in cottage cheese cream. As Table II illustrates, a blend of xanthan gum, locust bean gum, and guar gum can be used to develop the viscosity necessary for good cling, but the active gum concentration is low enough so that mouthfeel and texture do not suffer (9). The cream remains homogeneous and uniformly mixed with curd and does not separate with time.

Liquid Cattle Feed Supplements

Liquid feed supplements, although not food in the sense that they are consumed directly by humans, are nonetheless an important factor in the food supply. Liquid feed supplements are basically one or more nutrient materials supplied in a liquid vehicle such as water or molasses. The great majority are added to the dry feed of feedlot cattle and are shipped and stored in large tanks where, because of the formation of precipitates or flocculants or the addition of insoluble material, maintaining uniformity is difficult. It is especially important to maintain uniformity because vitamins and trace minerals which are added to the ration tend to adsorb on the solid flocculants that form and will not stay uniformly distributed unless sedimentation is prevented by the use of a suspending agent or continuous agitation. Figure 6 is a typical liquid feed supplement formulation of the high-molasses, phosphoric-acid type. Note the similarity to fertilizer. Figure 7 illustrates the functionality of xanthan gum as a suspending agent at room temperature, and Figure 8 illustrates its stability at 95°F (10). These figures illustrate two points: (1) xanthan gum is an excellent suspending agent because it possesses a yield point, and (2) suspending qualities of xanthan gum are not significantly diminished by elevated temperatures as occurs with liquid feed supplements in warm weather. In addition, because of its pseudoplasticity, xanthan gum facilitates pumping.

Table I FORMULAS AND RESULTS
OF EXPERIMENTAL PASTEURIZED PROCESSED CHEESE SPREADS

TRIAL NUMBER	LBG	GUAR GUM		XANTHAN GUM	BODY RESILIENCY	SLICING PROPERTIES	MOUTH FEEL	FLAVOR RELEASE	SANDWICH MELT
	% GUM CONTENT				ORGANOLEPTIC EVALUATIONS				
1A	0.2	--	--	--	Fair	Tacky	Lumpy	Fair	Excellent
2A	0.5	--	--	--	Fair	Tacky	Lumpy	Fair	Good
3A	0.8	--	--	--	Fair	Tacky	Lumpy	Fair	Good
4A	--	0.2	--	--	Good	Good	Lumpy	Fair	Good
5A	--	0.5	--	--	Excellent	Excellent	Lumpy	Fair	Excellent
6A	--	0.8	--	--	Excellent	Excellent	Lumpy	Fair	Excellent
7A	--	--	--	0.2	Poor	Tacky	Smooth	Excellent	Good
8A	--	--	--	0.5	Poor	Tacky	Smooth	Excellent	Excellent
9A	--	--	--	0.8	Fair	Tacky	Smooth	Excellent	Excellent
10A	0.06	--	--	0.14	Fair	Tacky	Lumpy	Fair	Good
11A	0.05	--	--	0.45	Fair	Tacky	Lumpy	Fair	Good
12A	0.15	--	--	0.35	Good	Tacky	Smooth	Good	Excellent
13A	0.25	--	--	0.25	Fair	Tacky	Lumpy	Fair	Good
14A	0.24	--	--	0.56	Good	Tacky	Smooth	Good	Excellent
15A	0.06	0.04		0.10	Good	Tacky	Smooth	Good	Good
16A	0.15	0.10	--	0.25	Good	Good	Smooth	Good	Excellent
17A	0.24	0.16	--	0.40	Excellent	Excellent	Smooth	Excellent	Excellent
18A*	0.15	0.10	--	0.25	Good	Good	Smooth	Good	Good

*Trial 18A contained 0.2 percent pimento solids or 12.50 pounds of drained pimentos per batch.

Table II Dressing Viscosities (cps) (Initial/24 hr.)

Percent Stabilizer:	0.05	0.10	0.15	0.20	0.30
Xanthan gum/ galactomannan blend	27/41	118/115	260/245	Not evaluated	
Xanthan gum	10/10	40/60	107/133	155/225	373/480
Guar gum	6/4	15/31	33/60	51/86	160/250
Blend 1	7/8	9/21	16/42	23/47	45/71
Blend 2	6/5	8/11	13/19	18/35	27/58
Blend 3	4/12	7/19	18/41	36/67	62/105
Blend 4	7/13	10/24	21/49	51/95	66/135
Locust bean gum	11/16	23/35	52/69	105/130	175/95

Cane Molasses, 79.5 Brix	67.5
Urea Liquor, 50%	20.9
Salt	5.0
Trace Minerals	0.2
Phosphoric Acid, 75%	6.4
	100.0

	%
Protein Equivalent	32
Phosphorus	1.5
Solids, Calculated	60

Figure 6. Liquid supplement formulation high molasses, range type

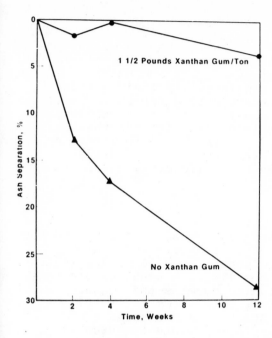

Figure 7.   *Ash separation vs. time, 67.5% molasses, room temperature*

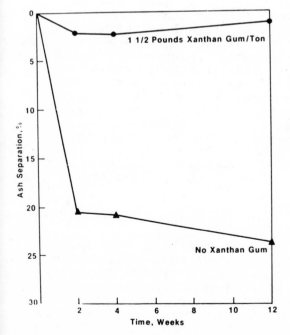

Figure 8.   *Ash separation vs. time, 67.5% molasses, 95°F*

Calf Milk Replacers

A related application is the use of xanthan gum for stabilizing calf milk replacers. Calf milk replacers may consist of dried whey and heat-processed soy beans or single-cell protein. A dry mix is added to water, stirred to disperse the solids, and fed to calves. If the feed is not stabilized, the insoluble solids will quickly sink and the short-term uniformity which is required will be lost. As little as 0.032 percent xanthan gum can provide enough stability to maintain uniformity, as Figures 9, 10, and 11 show (11).

Thermal Processing

Recent work by Cheng and Kovacs has explored the rheological properties of xanthan gum under high temperature and moderate shear found in most agitated commercial canning systems (12). A Fann 50C viscometer, an instrument which permits the measurement of viscosity at shear rates from 1.7 - 1075 sec^{-1}, temperatures to 500°F, and pressures to 1000 psi, was used. Figure 12 depicts the viscosity drop that xanthan gum undergoes at two arbitrarily selected shear rates, 170 sec^{-1} and 511 sec^{-1}, with and without NaCl. The well-known "flat" viscosity versus temperature curve is seen from ambient temperature to about 190°F with all solutions. The stability to thermal degradation imparted by an electrolyte can also be seen. Furthermore, Figure 9 illustrates that the electrolyte-stabilized solutions lose 98 percent of their viscosity at retort temperature (250°F) and recover about 80 percent of their original viscosity on cooling. Obviously, a thickener that is thin at retort temperature will facilitate heat transfer, thereby shortening the process time. This reduction in process time is important for increasing productivity and in the thermal processing of foods which are adversely affected by "overcooking".

Other Advances

Xanthan gum has also been used to stabilize frozen desserts and to stabilize toothpaste, where its rheological properties can be used to formulate a product that thins when squeezed from the tube but has the original consistency on the brush. A synergism with dextrin has also been used in denture

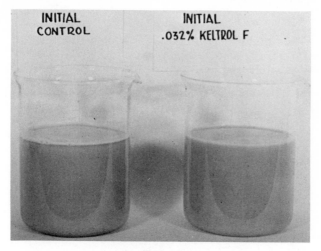

Figure 9

Figure 10

Figure 11

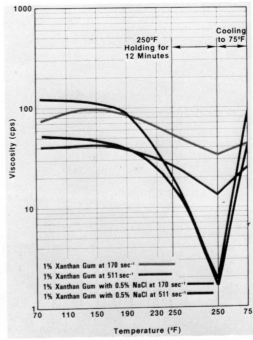

Figure 12. The effect of retorting at
shear rates of 170 sec⁻¹ and 511 sec⁻¹
on the viscosities xanthan gum with
and without electrolyte

adhesives and in a number of foods. Xanthan gum
alone and in combination with galactomannans shows
excellent promise in canned, gravy-type pet foods.
It has also been reported that xanthan gum shows
good potential as a gluten substitute in bread
(13, 14). Although this potential use is important
to the segment of the population who are allergic
to gluten, it has much broader implications as a
possible adjunct to low-quality wheat flour or
grains that are not suitable for baking.
   We have reviewed today some of the more recent
developments in xanthan gum usage. In view of the
unique properties it possesses as a result of its
function in nature, we can undoubtedly expect even
wider usage in the future.

Literature Cited

1.   "Xanthan Gum/KELTROL/KELZAN, A Natural Polysac-
     charide for Scientific Water Control," Kelco Co.,
     Second Edition.
2.   Sloneker, J. H., James A., Am. J. Chem. (1962) 40,
     2066-71.
3.   Siddiqni, Carbohyd. Res.   (1967) 4 (4), 284-91.
4.   Jansson, P. E., Keen, L., and Lindberg, B.,
     Carbohyd. Res. (1976) 45 (1), 275-82.
5.   Rees, D. A., Biophysical Society Winter Meeting,
     London, England (1973).
6.   Rees, D. A., Biochem. J. (1972) 126, 257-73.
7.   Dea, I. C. M., McKinnon, A. A. and Reese, D. A.
     (1972) J. Mol. Biol. 68 (1), 153-72.
8.   Kovacs, P., and Igoe, R. S. (1976) Food Product
     Development, in press.
9.   Kovacs, P. and Titlow, B. D. (1976) American
     Dairy Review 38 (4) 34J-34N.
10. Jackman, K. R., Randel, J. H. and Wintersdorff,
     P. (1976) A Unique Suspending Agent, National
     Feed Ingredients Association Meeting, Kansas City,
     Missouri (April 15).
11. Ibid.
12. Kovacs, P. and Cheng, H. (1976) Effects of
     High Temperatures and Shear Rates on Hydrocolloid
     Viscosities During Simulated Canning and HTST
     Processing Conditions. Unpublished report.
13. Kulp, K., Hepburn, F. N., and Lehmann, T. A.
     (1974) The Baker's Digest 48 (3), 34-37.
14. Christianson, D. D. Gardner, H. W., Warner, K.,
     Boundy, B. K. and Inglett, G. E. (1974) Food
     Technology 28 (6), 23-29.

19

Application of Xanthan Gum for Enhanced Oil Recovery

E. I. SANDVIK and J. M. MAERKER

Exxon Production Research Co., P.O. Box 2189, Houston, TX 77001

High molecular weight water soluble polymers find application in two different enhanced oil recovery processes. At present, the principal use is for an improved form of waterflooding in which polymers are used to increase the efficiency with which water can contact and displace reservoir oil. However, it is anticipated that polymer requirements for processes of this type will be overshadowed by the quantity needed to provide mobility control for future micellar-polymer projects. The latter processes have potential for producing oil that is unrecoverable by polymer augmented waterflooding. In both applications--polymer waterflooding and micellar-polymer flooding--the function of polymer is to reduce the mobility of injected water.

Mobility is defined as the flow capacity of a rock-fluid system, or the volumetric flow rate per unit area achieved with a given pressure gradient (Figure 1). It is usually expressed as effective rock permeability divided by fluid viscosity, and the common petroleum reservoir engineering units are darcies (or millidarcies) per centipoise. As will be discussed later, polymer can reduce mobility by decreasing effective rock permeability and by increasing effective fluid viscosity.

Effects of mobility and mobility ratio on the efficiency of reservoir displacements may be illustrated with Hele-Shaw models (1) which give a simplified portrayal of areal displacement efficiency in a reservoir element. These models commonly consist of two square glass plates that are separated and sealed at the edges by a thin spacing gasket and have provisions to inject and withdraw fluid at opposite corners to simulate injection and production wells. Mobilities of displacing and displaced fluids

may be varied by changing fluid viscosities. The effect of
mobility ratio on displacement efficiency is illustrated by the
Hele-Shaw model results depicted in Figure 2. The models are
initially oil filled, and water is injected in the lower left
corner. When the mobility of displacing water (unshaded) equals
the mobility of displaced oil (shaded), about three-fourths of
the oil is produced before water arrives at the production well
(Figure 2a). After water breakthrough, production increases in
water content with continued throughput. At a favorable viscosity ratio of 0.03, about nine-tenths of the oil is produced
before water breakthrough (Figure 2b). An unfavorable viscosity
ratio of 30 causes an obviously unstable displacement to occur,
and only about one-third of the resident oil is produced at water
breakthrough (Figure 2c).

  These experiments illustrate the importance of maintaining
as favorable a mobility ratio as possible during displacements.
The displacement illustrated in Figure 2c can be made to look
like the one in Figure 2a by increasing the water-phase viscosity
thirtyfold, and this would substantially improve oil recovery.
Similar results can be expected in actual reservoir situations.
Consequently, mobility control polymers are used in recovery
processes to reduce the mobility of injected water and increase
process efficiency.

  Two basic types of polymers--xanthan gums and partially
hydrolyzed polyacrylamides--constitute the large majority of
those currently used in enhanced recovery. At present, polyacrylamides strongly dominate polymer waterflooding applications,
while xanthan gums play a very minor role. Substantial increases
in xanthan gum use can be expected, however, as micellar-polymer
processes are further tested and then applied in larger-scale
applications. This paper is devoted primarily to a comparison of
the performance for these two polymer types in laboratory evalua-
tions and actual reservoir use. Hopefully, this comparison will
serve to pinpoint specific assets or liabilities and provide
guidelines for needed improvements.

Mobility Reduction

  As mentioned earlier, dilute polymer solutions work in two
ways to reduce water mobility in porous media: 1) by increasing
viscosity and 2) by decreasing permeability. Different polymers
depend on these two mechanisms in varying degrees. However, both
mechanisms are influenced by molecular weight, molecular weight
distribution, salinity, flow rate and permeability. In the
concentration range usually considered for enhanced oil recovery
applications -- 200 to 1500 ppm -- and in water salinities
normally encountered in reservoirs, Xanthan gums generally
exhibit higher viscosity and a lower sensitivity of viscosity to
salinity changes than partially hydrolyzed polyacrylamides.
Figure 3 shows viscosity-concentration behavior for several

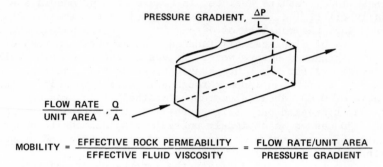

PRESSURE GRADIENT, $\frac{\Delta P}{L}$

$\frac{FLOW\ RATE}{UNIT\ AREA}$, $\frac{Q}{A}$

$MOBILITY = \frac{EFFECTIVE\ ROCK\ PERMEABILITY}{EFFECTIVE\ FLUID\ VISCOSITY} = \frac{FLOW\ RATE/UNIT\ AREA}{PRESSURE\ GRADIENT}$

Figure 1.   Definition of mobility

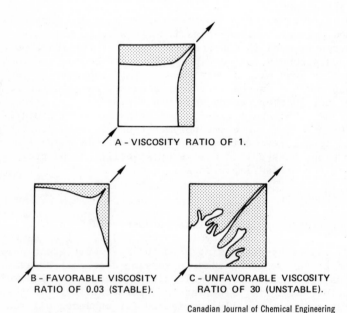

A - VISCOSITY RATIO OF 1.

B - FAVORABLE VISCOSITY
RATIO OF 0.03 (STABLE).

C - UNFAVORABLE VISCOSITY
RATIO OF 30 (UNSTABLE).

Canadian Journal of Chemical Engineering

Figure 2.   Displacements in Hele–Shaw model at different viscosity ratios (1)

polymers in two brines: 1% NaCl and a synthetic reservoir brine
containing calcium and magnesium. Each xanthan sample represents
a different commercial source, and, although they differ significantly from each other in viscosity, each exhibits little salt
sensitivity. The polyacrylamide, on the other hand, gives significant viscosity differences for the two brines used.

Many investigators (2-20) have observed permeability reduction with polyacrylamide solutions by flushing a polymerflooded
sandstone core or sand pack with brine and comparing the flushed,
brine mobility with that of brine prior to polymer. The ratio of
initial brine mobility to brine mobility after injection of a
polymer bank has been called a residual resistance factor, but
this factor is not always a good measure of permeability reduction during polymer flow (9). Two possible mechanisms can be
responsible for permeability reduction: 1) adsorption of polymer molecules on main-flow-channel walls which reduces cross-
sectional area available for flow, and 2) entrapment of polymer
molecules in narrow pore constrictions which partially shuts-off
a portion of the interconnected pore network. Interested readers
may gain an appreciation for each mechanism by comparing the
works of Thomas (19) and Domingues and Willhite (20). The
degree of permeability reduction varies inversely with original
brine permeability (2-4, 10, 21). This relationship is readily
understandable by recognizing that an adsorbed polymer molecule
of given size will cause a greater percentage reduction of cross-
sectional area in a small diameter pore (lower permeability) than
in a larger pore (higher permeability).

Xanthan gum solutions cause very little reduction of permeability in porous media (4, 19). As a result, mobility control
design for a secondary (polymer waterflood) or tertiary
(micellar-polymer flood) oil recovery process is simplified (7),
but the very real advantage of continued injection at a reduced
mobility is lost for brine injected behind a xanthan gum polymer
bank. Figure 4 compares resistance factors as a function of
throughput in one-foot Berea sandstone cores for a 600-ppm polyacrylamide solution and a 750-ppm xanthan gum solution. Under
these test conditions, steady-state mobility reduction during
polymer flow and the residual resistance factor (permeability
reduction) after brine flow are larger for the polyacrylamide,
even though it has less viscosity. However, it must be noted
that, because of the previously mentioned dependence of permeability reduction by polyacrylamides on initial brine permeability, much less mobility reduction would be expected if the
polyacrylamide test of Figure 4 had been conducted in 500-md,
rather than 100-md, sandstone. In the case of xanthan gum,
little change in mobility reduction would be expected with
changes in initial brine permeability.

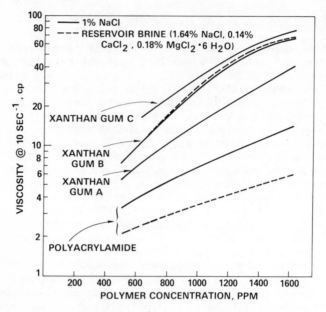

Figure 3.   Comparison of polymer viscosity–concentration behavior

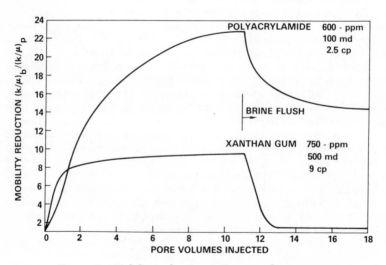

Figure 4.   Mobility reductions in Berea sandstone cores

Polymer Retention

Polymer retention in porous rock delays polymer bank arrival at a producing well and increases the quantity of polymer required to provide mobility control throughout an oil reservoir. Consequently, this loss of polymer to the formation must not be excessive. The two mechanisms mentioned above for permeability reduction -- chemical adsorption and physical entrapment -- are also two ways in which polymer molecules are removed from solution when flowing through porous media. Although polymer retention and permeability reduction are definitely interrelated, no one has yet determined the separate contributions from adsorption and physical entrapment for a given polymer-rock system. Several workers (6, 10, 19, 20,) have published data indicating that physical entrapment is the more important mechanism for polyacrylamides in water-wet sandstone cores or sand packs. Polyacrylamides can easily lose 300 to 400 pounds per acre-foot in consolidated sandstone. As with permeability reduction discussed earlier, this loss is also an inverse function of permeability (21). Xanthan gums exhibit less retention--on the order of 150 to 300 pounds per acre-foot. Further work is necessary to assess relative importance of adsorption and entrapment for xanthan gum solutions.

Inaccessible Pore Volume

It has been shown (19, 22) that molecular size for polymers of interest here can exceed the diameters of some of the smaller pores in natural porous media (16). This implies that a portion of the interconnected pore volume is inaccessible to polymer molecules. When polymer solutions flow through porous media, the result is an acceleration of polymer through larger pores relative to simultaneously injected solvent (23), in a manner reminiscent of gel permeation chromatography. This effect is illustrated by Figure 5, which shows effluent concentration response to a pulse of polyacrylamide and a tracer injected simultaneously into a sandstone core that had previously been contacted with a higher polymer concentration to satisfy retention. The polymer pulse, therefore, is not delayed by retention and breaks through about 22% of a pore volume early because 22% of the pore space is inaccessible to polymer molecules.

Inaccessible pore volume does not require pore constrictions too small for polymer molecules to pass; Thomas (19) has shown that bridging of adsorbed molecules in constant-radius capillaries with diameters less than four times the average molecular diameter can lead to redirection of subsequent polymer into larger flow channels. Polymer may also be accelerated relative to its solvent by a mechanism termed hydrodynamic chromatography (24) whereby the mean velocity of a particle in flowing fluid is a reflection of the pore velocity profile. Because of the size

of polymer molecules, their centers are excluded from the slowest
streamlines closest to pore walls, and they move faster than the
average solvent flow rate.

Transient Flow Behavior

Several investigators (9, 20, 25), have observed that, in
porous media, a steady-state equilibrium exists between retained
and flowing polymer. Interruptions or changes in flow rate can
perturb this steady state, resulting in transients in both retention and solution concentration. Figure 6 illustrates this
behavior for xanthan gum. Effluent concentration and mobility
reduction (resistance factor) are plotted versus pore volumes
injected for a 500-ppm xanthan gum solution with 2 percent NaCl
in a 6-inch, 121-md Berea core. At position A, flow was inter-
rupted for 16 hours and then resumed at the same pressure drop.
This resulted in sharp increases in both effluent concentration
and the degree of mobility reduction relative to previous steady-
state conditions. This result may be explained with the same
mechanistic considerations treated earlier--that is, under a
positive pressure gradient polymer molecules become packed into
pore cavities that have downstream outlets so constricted that
molecules cannot pass through. This contributes to permeability
reduction. Cessation of flow eliminates hydrodynamic drag and
permits the molecules to assume relaxed configurations. Molecular diffusion is then able to reduce the concentration gradients existing between cavities with restricted flow and main
channels. When flow is resumed, the increased concentration of
flowing polymer increases viscosity and, hence, also increases
mobility reduction. Permeability may also increase, but evidently this is overwhelmed by the attendant increase in viscosity. Subsequently, polymer trapping recurs and decreases the
effluent concentration below the injected value. This, in turn,
lowers in situ solution viscosity and mobility reduction. When
all trapping sites are once again saturated, the system returns
to its initial steady state.
      Positions B and C in Figure 6 indicate where pressure drop
across the core was increased without interrupting the flow. In
these cases additional polymer is immediately retained, lowering
both effluent concentration and the amount of mobility reduction.
The minima and asymptotic approaches to steady state with continued injection occur as before, but a lower equilibrium
mobility reduction results for each increase in pressure drop.
This is attributed to lower polymer solution viscosities at
higher shear rates (pseudoplastic, non-Newtonian behavior).
Here again, the incremental reduction of permeability associated
with additional polymer retention, which opposes the effect of
viscosity on mobility reduction, is dominated by the viscosity
contribution to mobility reduction.
      Comparing the behavior outlined above for xanthan solutions

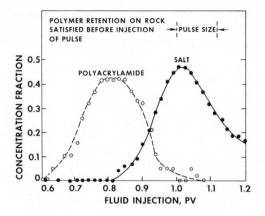

Society of Petroleum Engineers Journal

Figure 5.   Inaccessible pore volume (23)

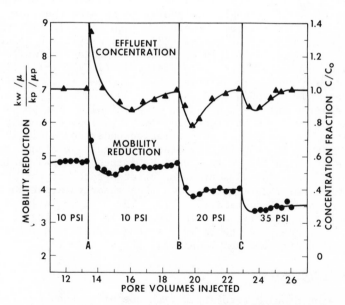

Journal of Petroleum Technology

Figure 6.   Effluent concentration and mobility reduction profiles vs. pore volumes injected for a 500-ppm xanthan gum solution with 2% NaCl (9)

with similar transient experiments for polyacrylamide solutions
(20) again shows a basic difference in porous media behavior,
which may be attributed to molecular conformation differences.
Resumption of polyacrylamide solution flow after a short interruption results in a mobility reduction decrease. This occurs,
apparently, because the effect of increased permeability resulting from dislodging molecules that had been trapped in pore constrictions is greater than the simultaneous viscosity increase
attributable to a higher flowing concentration. Polyacrylamide
solutions also show higher steady-state mobility reductions following increases in flow rate, but it is not clear whether this
effect is due mainly to reduced permeabilities from higher retention levels or higher extensional viscosities resulting from the
viscoelastic nature of polyacrylamide solutions (26). In contrast, xanthan gum solutions are relatively inelastic.

Mechanical Degradation

One critical problem arising from injection of polymer solutions into oil reservoirs is the possibility of imposing fluid
stresses large enough to rupture molecules and reduce molecular
weight. Because of the radial-flow nature of injection wells,
fluids entering a formation at typical flow rates are subjected
to very high fluxes at the sand face. These large fluxes and
the converging-diverging nature of flow channels in porous media
cause sections of entangled molecules to be stretched very
rapidly, and some molecules rupture before entanglements can
rearrange to relieve the stress (27). Polyacrylamide solutions
are very susceptible to this mechanical degradation, while
xanthan gum solutions are quite resistant (7, 27, 28). Figures 7
and 8 compare shear viscosities vs shear rate before and after
high-shear flow through bead packs for 300-ppm solutions of a
polyacrylamide and a xanthan gum, respectively. The polyacrylamide solution shows an eightfold viscosity loss following bead
pack flow, whereas the xanthan solution undergoes negligible viscosity loss after experiencing order of magnitude higher shear
rates in the bead pack. It should be pointed out that velocity
gradient in the flow direction, or stretch rate, appears to be a
better measure of deformation rate than shear rate for correlation of mechanical degradation (27). The maximum in apparent
viscosity for polyacrylamide in Figure 7 is due to its viscoelastic character, and there is a strong correlation between
viscoelastic stresses and mechanical degradation (27).

Injectivity Behavior

A major problem associated with use of xanthan gum for
mobility control applications has been poor injectivity behavior.
This problem may be illustrated by injectivity tests (28) conducted in the Coalingua Field, California. During injection of

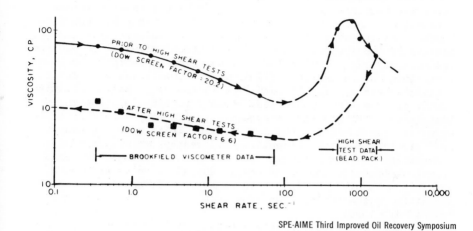

SPE-AIME Third Improved Oil Recovery Symposium

Figure 7.   Viscosity vs. shear rate for 600-ppm polyacrylamide in 300-ppm NaCl brine (7)

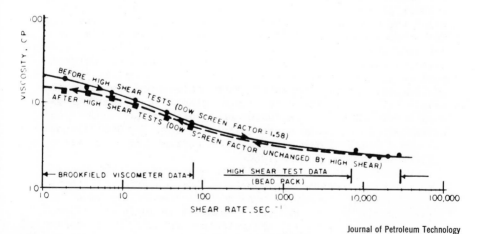

Journal of Petroleum Technology

Figure 8.   Viscosity vs. shear rate for 600-ppm xanthan gum in 300-ppm NaCl brine (28)

water and xanthan gum solutions, flow rates and pressures were
monitored to determine injectivities of the various fluids.
Injection was through casing perforated with 4 shots-per-foot
into either one or two sands at depths of 1600 to 1800 feet sub-
surface. Softened fresh water was used for polymer hydration to
yield a 6000-ppm concentrate and for further dilution to 300 ppm
prior to cartridge filtration and injection. Figure 9 shows
injectivity as a function of injected volume. During injection
of xanthan gum (points "L" and "P", Figure 9), substantial injectivity decreases were observed. Upon returning to brine injection (points "M" and "Q", Figure 9), injectivity levels remained
significantly below pre-polymer values. This indicated near-
wellbore plugging, which could be removed by two well cleanup
procedures: either a simple back wash or a treatment designed
to remove bacteria and other debris as indicated by points "N"
and "R" in Figure 9. Clearly there are conditions of use where
injection of xanthan gum can result in undesirable plugging and
injectivity loss.

Injectivity problems with xanthan may be attributed to
three primary factors: water quality, polymer composition and
injection well configuration. In general, water quality requirements are more stringent for polymer than for plain water injection. In the Coalingua injection test, it was concluded that
finely divided solids in the injection source water were being
flocculated by polymer and contributed to plugging problems (28).
In addition, any species present in the injection water that can
crosslink xanthan, such as ferric iron or borate ion, should be
avoided.

Composition of the polymer as commercially supplied is
another major cause of xanthan injectivity problems. Again
using the Coalingua tests as an example, it was concluded (28)
that the xanthan contained about 11 weight percent cellular
debris. This cellular material and unhydrated polymer "gels"
were believed responsible for most of the plugging. Several
publications and patents deal with xanthan injectivity problems
and methods for improvement. These include techniques to flocculate cellular debris onto clay or other solids that enable
easier removal by filtration or sedimentation (29, 30), procedures for diatomaceous earth filtration (28, 29, 30, 31) and
methods for dissolving protienaceous debris through alkaline
(32) or enzymatic (33) action.

While all of these clarification procedures probably do
improve polymer-solution properties, some question exists as to
the validity of procedures used to evaluate the degree of
improvement. Both of the tests commonly employed--Millipore
filter tests and rock injection tests--are capable of ranking
polymer solutions on a relative basis, but tests reported in the
literature do not begin to approach the injection levels experienced in actual wells. Since the plugging usually observed with
xanthan gum solutions is a near-surface or sandface phenomenon,

it is reasonable to scale injection on the basis of volume
injected per unit area of sandface, as is commonly done in filtration studies. This means that the type of injection well
completion is an important factor, since completion type will
determine sand exposure. The following table shows injection
values for several well completions commonly employed. Injection
values can range about three orders of magnitude depending on the

SCALED INJECTION VALUES FOR VARIOUS WELL COMPLETIONS
ASSUMING 10 BBL/DAY-FT INJECTION

Well Completion	Scaled Injection, ml/cm^2-day
4 Collapsed Perforations/Ft	3.1×10^5
2 Perforations/Ft	1.3×10^4
6" Open Hole	1.1×10^3
18" Underreamed and Gravel Packed	3.6×10^2
"Typical" Laboratory	< 100

type of well completion. The collapsed perforation case is most
unfavorable from an injectivity standpoint (this is probably the
situation for the Coalingua tests described earlier). The most
favorable situation would be one where the well is enlarged
throughout the reservoir interval and packed with coarse sand or
gravel behind a screen. All of these well completions have
scaled injection values substantially above those generally
reported for laboratory tests. In other words, published test
data are short by a factor of up to a thousand of simulating
even one day's injection at a rate of 10 bbl/day/ft of interval.
    In an effort to investigate injectivity more realistically,
we have conducted injection tests in Berea cores at higher
throughput levels. Tests were conducted in one-half-inch square
by one-foot-long sandstone cores fitted with pressure taps as
shown in Figure 10. Fluids were pumped through each core at
constant rate, and mobilities, or flow capacities, of the
various sections were determined as a function of throughput.
With an appropriate experimental setup, brine could be injected
for sustained periods without plugging (Figure 11). Injection
of sheared, but unfiltered, polymer solution caused rapid plugging as shown in Figure 12. Filtration of the polymer solution
after shear improved injectivity (Figure 13) but certainly did
not eliminate plugging tendencies. Results of these laboratory
tests may be combined with a model for radial flow of a non-
Newtonian fluid in porous media (28) to calculate response of a
hypothetical injection well. Figure 14 shows pressure drop
between the wellbore and a point 100 ft into the formation
plotted as a function of injection time at 10 bbl/day/ft. If a
non-plugging polymer solution is injected, there will be some
pressure increase due to mobility reduction in the growing polymer bank. This is the desired pressure response. Injection of

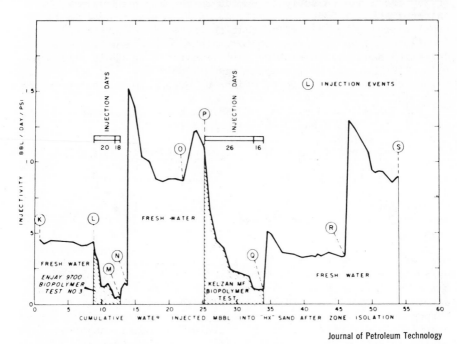

Figure 9.   Test-well biopolymer injectivity tests, May 8, 1971 to February 28, 1972 (28)

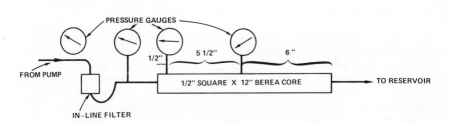

Figure 10.   Injectivity core test schematic

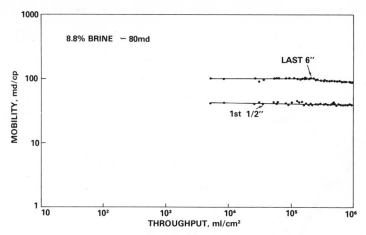

Figure 11.   Injectivity core test—8.8% brine

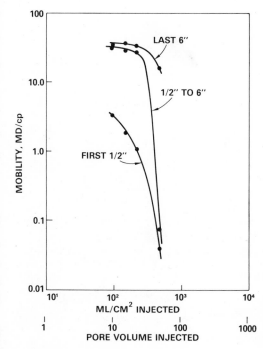

Figure 12.   Injectivity core test for 600-ppm sheared but unfiltered xanthan gum in 8.8% filtered brine—80 MD Berea core

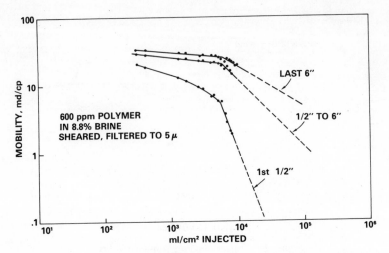

Figure 13.   Sheared and filtered xanthan gum injectivity core test

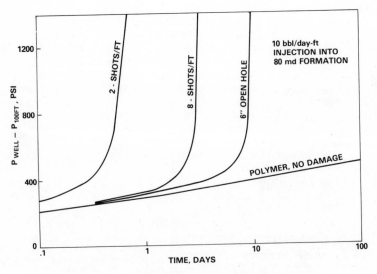

Figure 14.   Predicted injection well pressure response

a xanthan gum solution sheared and filtered to 5 microns will
give one of the upper curves depending on well completion.
Again, this calculation is based on laboratory data, but it is
reasonably consistent with field experience. Clearly, injection-
well plugging can be expected for some field situations, and
there is substantial room for improvement in the xanthan polymers
currently in use.

Compatibility

Injectivity problems experienced with xanthan gums can be
regarded as one form of compatibility problem--something in the
polymer solution is too big to enter the formation. Other compatibility considerations when polymers are used in enhanced
recovery processes include: interactions with reservoir brines,
sensitivity to degradation by thermal, chemical, or microbial
action, and, in some cases, interaction between polymer and
micellar fluids.

Native formation brines vary in salinity from potable fresh
water to greater than 20% dissolved inorganic solids. Figure 15
shows how intrinsic viscosity of a particular partially hydrolyzed polyacrylamide varies with brine concentration and composition. In sodium chloride alone, behavior is typical of most
"flexible" polyelectrolytes, i.e. intrinsic viscosity varies
linearly with ionic strength to the minus one-half power (34).
Greater intrinsic viscosity reductions are observed with calcium
containing brines--probably due to crosslinking. In general, as
water salinity increases, polyacrylamide effectiveness decreases
due to decreases in viscosity and permeability reduction (2, 4)
and an increase in mechanical degradation (27). There is
obvious incentive for using the lowest salinity injection water
feasible for any specific application. Xanthan shows little
change in fluid properties with salinity; this apparently is due
to the more rigid nature of the hydrated molecule.

Most current applications of polymers are in reservoirs
having temperatures of 160°F (∿70°C) or below. In this range,
thermal degradation is not considered to be a serious problem.
It is anticipated that future applications will be at temperatures up to at least 250°F (∿125°C). At some point, thermal
degradation will be a concern for both polyacrylamides and
xanthan, but published work has not yet clearly defined this
point.

Substantial chemical degradation at relatively low temperatures has been observed with polyacrylamides under certain conditions. If both dissolved oxygen and redox catalysts--for
example, dissolved iron--are present, rapid degradation occurs
(35, 36). Generally this is avoided in laboratory work by
excluding metal from the experimental system. In field use,
oxygen is scavenged from injection water before polymer addition.
Figure 16 shows the effect of sodium hydrosulfite addition on

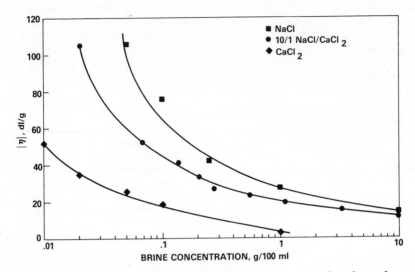

Figure 15. Intrinsic viscosity of partially hydrolyzed polyacrylamide vs. brine concentration

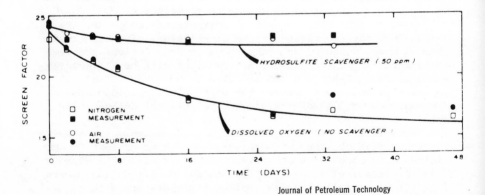

Journal of Petroleum Technology

Figure 16. Effect of hydrosulfite on polymer degradation (36)

polymer quality. Slow but substantial degradation occurs with
dissolved oxygen present; much less occurs when oxygen is scavenged by hydrosulfite.

In the case of xanthan gum, chemical degradation is not
considered to be a problem, but biodegradation can cause severe
viscosity loss and plugging by microbial slimes. Biodegradation
has been eliminated by use of chlorophenolates, formaldehyde, or
adjustment of injection water pH to a high value.

Another compatibility aspect for mobility control polymers
arises through interactions between a polymer drive bank and the
surfactant slug employed in micellar-polymer processes. These
processes depend on the ability of injected surfactants to lower
interfacial tension between trapped residual oil and the displacing fluid. Because of high chemical costs, only small surfactant banks can be used. Frequently, these surfactant solu-
tions--or microemulsion formulations--have fairly high viscosities (ten to twenty times water viscosity). If ordinary brine
were used to displace a microemulsion slug, brine would finger
into the slug because of the unfavorable mobility ratio, dilute
the surfactant, and render it ineffective. Consequently, it is
important to follow a surfactant slug with a mobility buffer.
Polymer solutions having mobilities equal to or less than the
surfactant slug have been the primary choice for this essential
task.

It has been observed that polymer retention decreases to
nearly insignificant levels when a polymer bank follows a surfactant slug (37). As a result of this, and because of polymer
inaccessible pore volume, polymer molecules from a mobility
buffer bank can invade a surfactant slug. Phase behavior
studies have shown (38) that presence of xanthan gum molecules
in a microemulsion consisting of brine, IPA and a petroleum sulfonate can cause the mixture to separate into two phases. This
has been called sulfonate-polymer interaction.

Sulfonate-polymer interaction can be very detrimental to
oil recovery operations by contributing to surfactant loss.
Figure 17 shows a plot of produced sulfonate, polymer and tracer
concentrations as fractions of injected values vs pore volumes
produced in an 8-foot-Berea-core tertiary flood. Sulfonate concentration begins to drop when polymer first appears in the
effluent. The shaded area between the sulfonate and its tracer
concentration profiles represents sulfonate loss due to trapping
of a second phase. Since the sulfonate tracer concentration
does not drop until the polymer tracer appears in the effluent,
polymer apparently has moved into the surfactant slug. In some
instances, pressure gradients required to maintain realistic oilfield flow rates become excessive. Minimizing this detrimental
phase behavior by lowering drive water salinity and increasing
cosurfactant/surfactant ratio (38) may reduce sulfonate trapping
from sulfonate-polymer interaction and improve tertiary oil
recovery.

Field Application

Previous discussion has been concerned primarily with
laboratory measurements of the many aspects of mobility-control
polymer behavior. Because of the variety of interactions with
porous media and the complex nature of petroleum reservoirs, the
real test of a polymer and a process is actual field performance.
Figure 18 illustrates typical steps and interactions involved in
evaluation of a proposed field project. For each polymer under
consideration, laboratory measurements are made of the various
aspects of polymer behavior described earlier. This is done
using rock, water, and oil from the subject reservoir and specif-
ically includes retention and mobility reduction as a function of
polymer concentration, rock permeability, etc. Measurements or
estimates of injectivity behavior and degradation level are also
desirable. This polymer information is combined with best
available reservoir description to construct a computer model of
the reservoir and process. Calculations from this model may be
used for optimization and prediction of polymer flood and waterflood performance. These performance estimates are then used to
provide an economic evaluation of a polymer project relative to
an ordinary waterflood project. One way of comparing behavior
is by a plot of produced water-oil-ratio vs cumulative oil
production (Figure 19). Each project is terminated when the
water-oil-ratio becomes high enough that continued operation is
no longer economic (in this case, 96% water cut, for example).
It is interesting to note the relatively small amount of additional oil recovered by the polymer flood when compared to the
waterflood base case. Even using polymer, the amount of oil
left at project termination could easily be about the same as
the total polymer flood recovery--about 300 bbl/Ac-Ft of reservoir. This remaining oil is then a potential candidate for
micellar-polymer flooding.

The preceeding discussion gives a very rough estimate of
how one evaluates a polymer waterflood project. An obvious
question at this point is--"where does xanthan gum fit in?" The
answer is that xanthan gum has played a very minor role. The
large majority of polymer waterflood projects, past and present,
are using some form of polyacrylamide. This does not mean,
however, that xanthan gum will continue to play a minor role in
the future. Improvements in the commercial product should
enable xanthan gums to capture a greater fraction of polymer
waterflooding applications. On the other hand, forecasts for
enhanced recovery agree that polymer flooding is expected to be
relatively minor compared to micellar-polymer flooding. Figure
20 shows the distribution of current and planned micellar-
polymer projects in the U.S. (40). It has not been possible to
determine what polymer is being considered or used for every
project, but, based on the polymer chosen for a significant
fraction of these tests, xanthan gum will be used in

1. 1.4 PV PETROLEUM SULFONATE IN 92% 0.23N NaCl
2. 1.5 PV 700 PPM XANTHAN GUM IN 0.05N NaCl
3. 2.0 PV 0.05N NaCl

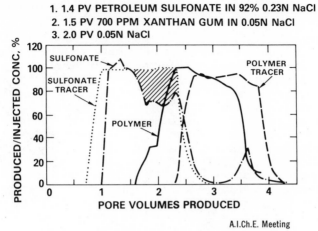

A.I.Ch.E. Meeting

Figure 17.   8-ft Berea core tertiary flood—110°F (38)

POLYMER FLOOD EVALUATION

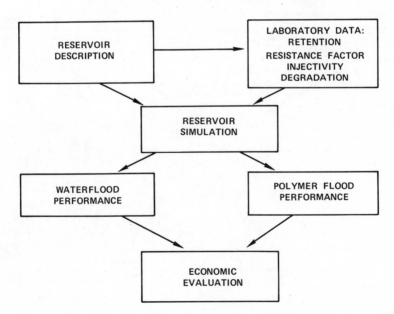

Figure 18.   Steps for evaluating a proposed field project

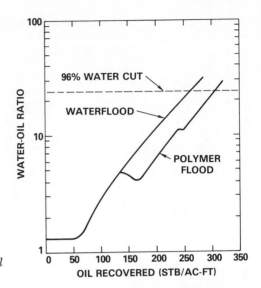

Figure 19. Polymer and waterflood oil production response

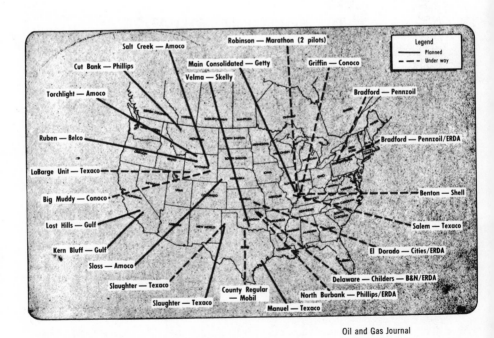

Oil and Gas Journal

Figure 20. Micellar/surfactant projects testing U.S. reservoirs (40)

approximately one-third to one-half. If this ratio holds for future full-scale projects, consumption of xanthan gum in enhanced recovery projects is likely to be substantial.

Literature Cited

1. Mungan, N., Can. J. Chem. Eng. 49, 32-37 (1971).
2. Smith, F. W., J. Pet. Tech. (Feb. 1970) 148-156.
3. Jennings, R. R., Rogers, J. H., and West, T. J., J. Pet. Tech. (March 1971) 391-401; Trans., AIME, 251 (1971).
4. Hirasaki, G. J. and Pope, G. A., Soc. Pet. Eng. J. (Aug. 1974) 337-346.
5. Mungan, N., Smith, F. W., and Thompson, J. L., J. Pet. Tech. (Sept. 1966) 1143-1150; Trans., AIME, 237 (1966).
6. Szabo, M. T., Soc. Pet. Eng. J. (Aug. 1975) 323-337; Trans., AIME, 259 (1975).
7. Hill, H. J., Brew, J. R., Claridge, E. L., Hite, J. R., and Pope, G. A., paper SPE 4748 presented at the SPE-AIME Third Improved Oil Recovery Symposium, Tulsa, Okla. (April 22-24, 1974).
8. Chauveteau, G. and Kohler, N., Paper SPE 4745 presented at the SPE-AIME Improved Oil Recovery Symposium, Tulsa (April 22-24, 1974).
9. Maerker, J. M., J. Pet. Tech. (Nov. 1973) 1307-1308.
10. Gogarty, W. B., Soc. Pet. Eng. J. (June 1967) 161-173; Trans., AIME, 240 (1967).
11. Burcik, E. J., Prod. Monthly (June 1965) 29.
12. Burcik, E. J. and Walrond, K. W., Prod. Monthly (Sept. 1968) 12-14.
13. Ershaghi, I. and Handy, L. L., paper SPE 3683 presented at 42nd Annual California Regional Mtg. of SPE of AIME, Los Angeles (Nov. 4-5, 1971).
14. Knight, B. L., U.S. Patent 3,724,545. (April 3, 1973).
15. Norton, C. J. and Falk, D. O., U.S. Patent 3,743,018 (July 3, 1973).
16. Thakur, G. C., paper SPE 4956 prepared for the Permian Basin Oil Recovery Conference of the SPE of AIME, Midland, Texas (March 11-12, 1974).
17. Martin, F. D., Paper SPE 5100 presented at the SPE-AIME 49th Annual Fall Meeting, Houston, Texas (Oct. 6-9, 1974).
18. Sparlin, D., Paper SPE 5610 presented at the SPE-AIME 50th Annual Fall Meeting, Dallas (Sept. 28-Oct. 1, 1975).
19. Thomas, C. P., Soc. Pet. Eng. J. (June 1976) 130-136.
20. Domingues, J. G. and Willhite, G. P., Paper SPE 5835 presented at the SPE-AIME Improved Oil Recovery Symposium, Tulsa (March 22-24, 1976).
21. Vela, S., Peaceman, D. W., and Sandvik, E. I., Soc. Pet. Eng. J. (April 1976) 82-96.
22. Lynch, E. J. and MacWilliams, D. C., J. Pet. Tech. (Oct. 1969) 1247-1248.

23. Dawson, R. and Lantz, R. B., Soc. Pet. Eng. J. (Oct. 1972)
     448-452; Trans., AIME, 253 (1972).
24. Small, H., J. Colloid and Interface Science (July 1976)
     147-161.
25. Rhudy, J. S., Fullinwider, J. H., and Ver Steeg, D. J.,
     U.S. Patent 3,734,183 (May 22, 1973).
26. Marshall, R. J. and Metzner, A. B., Ind. and Eng. Chem.
     Fund., 6 (1967) 393-400.
27. Maerker, J. M., Soc. Pet. Eng. J. (Aug. 1975) 311-322;
     Trans., AIME, 261 (1975).
28. Tinker, G. E., Bowman, R. W., and Pope, G. A., J. Pet.
     Tech. (May 1976) 586-593.
29. Lipton, D., paper SPE 5099 presented at SPE-AIME 49th
     Annual Fall Meeting, Houston (Oct. 6-9, 1974).
30. Abode, M. K., U.S. Patent 3,711,463 (1973).
31. Yost, M. E. and Stokke, O. M., J. Pet. Tech. (Oct. 1975)
     1271-1272.
32. Patton, J. T., paper SPE 4670 presented at the SPE-AIME
     48th Annual Fall Meeting, Las Vegas, Nev. (Sept. 30-Oct. 3,
     1973).
33. Burnett, D. B., paper SPE 5372 presented at the 45th
     Annual California Regional Meeting of SPE-AIME, Ventura
     (April 2-4, 1975).
34. Smidsrod, O. and Haug, A., Biopolymers, 10, (1971) 1212-1227.
35. Pye, D. J., U.S. Patent 3,343,601 (Sept. 26, 1967).
36. Knight, B. L., J. Pet. Tech. (May 1973) 618-626.
37. Trushenski, S. P., Dauben, D. L., and Parrish, D. P.,
     Soc. Pet. Eng. J. (Dec. 1974) 633-642; Trans., AIME, 257
     (1974).
38. Trushenski, S. P. paper presented at A.I.Ch.E. meeting,
     Kansas City (April 1976).
39. Healy, R. N., Reed, R. L., and Stenmark, D. G., Soc. Pet.
     Eng. J. (June 1976) 147-160.
40. Oil and Gas J., 74, No. 14 (1976).

Production, Properties, and Application of Curdlan

TOKUYA HARADA

Institute of Scientific and Industrial Research, Osaka University,
Yamadakami, Suita-shi, Osaka, Japan (565)

Curdlan was produced in high yield by cultures of a newly isolated and improved mutant strain of Alcaligenes faecalis var. myxogenes. Curdlan forms a gel with specific properties and it should be a useful, new polymer not only as a food additive, but also for industrial purposes.

I. Findings

The history of the discovery of curdlan is interesting. In 1962, Harada and his colleagues made great efforts to obtain microorganisms which could utilize petrochemical materials. They isolated an organism from soil, capable of growing on medium containing 10% ethylene-glycol as the sole carbon source (1) and named it Alcaligenes faecalis var. myxogenes 10C3 (2, 3). They found that this organism produced a new β-glucan which contained about 10% succinic acid and named it succinoglucan (4, 5). The structure of the polysaccharide moiety of succinoglucan (6, 7) is shown below:

→G1cl→4G1cl→3G1cl→3G1cl→6G1cl→4G1cl→3G1cl→3G1cl→4G1cl→3Ga1l→4G1cl→

During investigations on the production of succinoglucan, one day they found that the culture medium did not become viscous and no succinoglucan was formed, but almost all the added glucose was consumed. They thought that some special compound(s), must have been produced instead of succinoglucan in the culture. So they examined the product and found that it was a neutral polysaccharide (8, 9). They named it curdlan in 1966 (10).

Curdlan is composed of β-1,3-glucosidic linkages. A mutant strain 10C3K was isolated from the stock culture 10C3, which produced only curdlan. Strain 10C3K is a spontaneous mutant and it has stable ability to produce the exocellular polysaccharide whereas the ability of strain 10C3 is unstable(11). Thus, by chance, they succeeded in obtaining a suitable organism for production of curdlan. Later Takeda Chemical Industries Ltd.isolated a uracil-less mutant of strain 10C3K named strain· 13140 as a better gel-forming β-1,3-glucan producer (12). The polymer from the strain, designated as polysaccharide 13140, is a kind of curdlan in a broad sense or a curdlan type polysaccharide.

II. Mutation

The detection of curdlan using Aniline Blue was tested using strain 10C3 and its mutant strains 10C3k and 22 as shown in Figure 1.(13). The culture medium used in this plate, consisted of 1% glucose, 0.5% yeast extract, 0.005% Aniline Blue and 2% agar. The middle colony is that of 10C3. The surrouding-clear zone is due to the formation of succinoglucan which is a soluble, viscous polymer. Succinoglucan does not stain with Aniline Blue. Curdlan can form a complex with this dye which is blue. The rate of interaction of the polymers with Aniline Blue was shown by Nakanishi and his colleagues to be proportional to their concentrations and degrees of polymerization (14). The left colony is that of a spontaneous mutant of the parent strain which produces only curdlan. The complex of the polymer with the dye can easily be stripped off. The remaining cells do not stain with the dye. The right colony is that of mutant strain 22, derived from 10C3 by treatment with N-methyl-N'-nitro-N-nitrosoguanidine . This strain produces only succinoglucan.
Mutation of strain 10C3 to strains staining with Aniline Blue was also induced by treatment with mutagens such as NTG, and ethylmethane-sulfonate and ultraviolet light, but not by treatment with Mitomycin C, ethidium bromide or Acridine Orange which are reagents causing elimination of plasmids (Table 1) (11). Experiments on transfer of genes concerned with production of succinoglucan and(or) curdlan between different mutant strains have not been successful. Thus, a plasmid may not be directly involved in the production of the polysaccharides.

III. Structure

Curdlan is composed of β-1,3-glucosidic linkages ($[\alpha]_D$ +18°, 1N NaOH) (10, 15). Saito and his colleagues (15) indicated the presence of two internal β-1,6-glucosidic linkages in original curdlan ($\overline{DP}$n 455) while Ebata (16) detected one part of gentibiose to 360 parts of glucose in the hydrolyzate of the glucan by the action of exo-β-1,3-glucanase, although Nakanishi and his colleagues could not detect any other glucosidic linkages besides β-1,3-glucosidic linkages in polysaccharide 13140 (12). Cellulose fiber, which is composed of β-1,4-glucosidic linkages, does not swell in the presence of water whereas curdlan swells in water and can form a resilient gel on heating. This is an important and interesting fact. Callose and pachyman are β-1,3-glucans which are largely composed of β-1,3-glucosidic linkages. Callose contains a little glucuronic acid (17). Pachyman has other gluco-

Table 1
Effects of Mutagens on Mutation of Strain 10C3 to Strains Staining
with Aniline Blue (11)

Mutagen	Concentration (per ml)	Growth inhibition (%)	Ratio of blue colonies to white colonies (%)
None			1.0×10^{-7}
N-Methyl-N'-nitro-N-nitroso-guanidine	30 μg	25	1.4×10^{-3}
Ethylmethane-sulfonate	6 mg	35	1.4×10^{-3}
5-Bromo-uracil	2 mg	0	1.0×10^{-7}
Ultraviolet light irradiation			2.0×10^{-3}

sidic linkages and does not form a resilient gel on heating (15).
Callose has been found in a variety of locations in the tissues
of higher plants, such as in sieve tubes, young tracheides, pollen,
root hairs, stem hairs and root endodermis. No other polysaccharides besides curdlan composed entirely of β-1,3-glucosidic
linkages have yet been found.

IV. Production

Now it has become possible to obtain heat-gelable β-1,3-glucan
easily from glucose and many carbon compounds. The yield of the
polymer from added glucose is about 50%. About 5 g of curdlan can
be produced from 10 g of glucose in 100 ml of simple defined
medium, if the pH is maintained at neutrality (9, 12, 18).
Curdlan can also be produced using a cell suspension in medium
containing only glucose and calcium carbonate (19). A pilot plant
for production of polysaccharide 13140 has been accomplished in
Takeda Chemical Industries Ltd.
Nakanishi and his colleagues examined the occurrence of
curdlan type polysaccharides in microorganisms, using the Aniline
Blue method (Table 2) (13). Four strains of Agrobacterium
radiobacter, one strain of Agrobacterium rhizogenes and a strain
of Agrobacterium sp. were found to produce curdlan type polysaccharides with water soluble β-glucans (11, 13). Spontaneous
mutant strains which produce principally curdlan-type polysaccharides in high yield were also induced from the respective
parent strains.

Table 2
Curdlan Type Polysaccharides (Curdlan in a Broad Sense)

Alcaligenes faecalis var. myxogenes 10C3K	Curdlan Polysaccharide 10C3K
Alcaligenes faecalis var. myxogenes IFO 13140	Polysaccharide 13140
Agrobacterium radiobacter	
IFO 12607	Polysaccharide 12607
IFO 12665	Polysaccharide 12665
IFO 13127	Polysaccharide 13127
IFO 13256	Polysaccharide 13256
Agrobacterium rhizogenes	
IFO 13259	Polysaccharide 13259
Agrobacterium sp. IFO 13660	Polysaccharide 13660

The structure of the polysaccharide moiety of a water soluble polymer from strain A. radiobacter IFO 12665 seems to be like that of succinoglucan because a specific β-glucanase, succinoglucan depolymerase from Flavobacterium sp. M64 (20), can attack the polymer to release oligosaccharide with similar Rf value to that of the product released from succinoglucan by the enzyme (21). Succinic acid may be not contained in the polymer.

V. Rheology

Excretion of curdlan as microfibrils from the cells of strain 10C3K, is seen by electron microscopy (Figure 2).
When a 2% suspension of curdlan is heated, it becomes clear at about 54°C and gel forms at higher temperature (22). Agar gel is formed when the sol of agar obtained by heating its suspension is cooled. This is a difference between curdlan and agar.
Figure 3 is a photograph of the gel of curdlan obtained by heating a 2% suspension at 90°C. The gel of curdlan is very elastic and resilient and does not break, whereas agar gel breaks when it is pressed between the fingers. The gel as seen in Figure 4 is easy to make using curdlan but it is not easy to make such gels using agar. It has also been found that curdlan forms a gel when an alkaline solution is dialyzed in a cellophan bag(S.Okamoto unpublished), when an aqueous solution of 0.2 - 0.63 M dimethylsulfoxide is cooled (23)or when calcium ions are added to a weakly alkaline solution (H. Kimura, unpublished).
Maeda and his colleagues investigated the effect of temperature on gel formation using a curdmeter. They heated 3% sus-

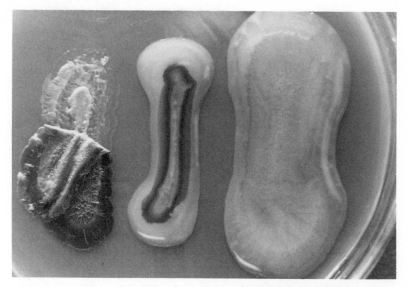

Journal of General and Applied Microbiology

Figure 1. *Photograph of colonies of strains (left to right) 10C3K, 10C3, and 22 grown on glucose–yeast extract medium containing water-soluble aniline blue (0.005%) (13)*

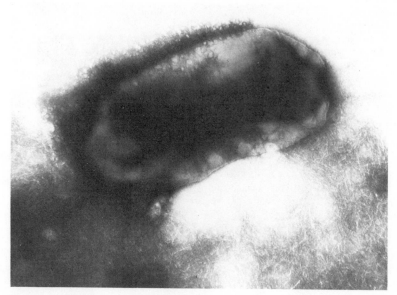

Figure 2. *Curdlan excreted from the cells of 10C3k as microfibrils, negatively stained with uranyl acetate*

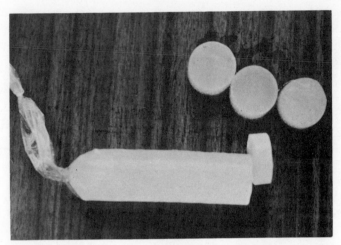

Figure 3.   Photograph of curdlan gel. Aqueous suspension (2%) of
this polymer was heated at 90°C for 10 min.

Figure 4.   Photograph of curdlan gel. Aqueous suspension (2%) of
this polymer was heated in special vessel at 90°C for a few min.

pensions of curdlan for 10 min and then measured the strength
of the resulting gels at 30°C (Figure 5). This curve is curious:
the gel strength is about the same between 60°C and 80°C and then
it increases from 80°C. The gel strength depends on the temperature but is independent of the incubation time at 70°C (22).

Urea breaks hydrogen bonds and its effect on gel formation
was investigated using a Shimadzu Microviscograph (Figure 6).
The starting temperature for gel formation decreased with increase
in the concentration of urea added. It is interesting that formation of gel in the second stage was not observed with above 5 M
urea. Thus, gel formation in the first stage seems to require the
breakage of hydrogen bonds whereas that in the second stage does
not. It is also interesting that the viscosity increased markedly
from 39°C to 20°C with 2 to 8 M urea when the temperature was
decreased. The formation of gel at low temperature may be due to
formation of hydrogen bonds.

Ethylene-glycol accelerates formation of hydrogen bonds and
its effect on gel formation was examined in the same way. The
results in Figure 7 show that it also decreased the starting
temperature for gel formation. However, in the presence of a high
concentration (5 M to 7 M) of ethylene-glycol, no gel was formed.
These results suggest that at the starting temperature for gel
formation some or all the hydrogen bonds must be broken.

The formation of gel of the polymer was investigated using a
Rotovisca Viscometer (Haake) by the members of Takeda Chemical
Industries Ltd.(24). The specific viscosities were determined
continuously as the temperature was raised to 60°C and then
decreased (Figure 8). From 54°C to 60°C swelling occurred due to
breakage of hydrogen bonds. On cooling the gel to about 40°C, the
viscosity rapidly increased and low-set gel was obtained (25). The
effect of temperature on transmittance was examined under the same
conditions (Figure 9). The transmittance increased on heating to
60°C and decreased on cooling from 60°C (25).

Figure 10 shows that the specific viscosity also increased to
some extent on cooling from 85°C with formation of high-set gel
(22, 24). As shown in Figure 11, the transmittance decreased
gradually with increase in temperature from 60°C to 100°C (25).
This was probably due to formation of hydrophobic bonds during
formation of cross links. Acetone powders were prepared from gels
formed by heating at 60°C, 70°C and 90°C. The two formers formed
similar gels to that derived from the original polymer, but a
powder from gel heated at 90°C did not. This indicates that gels
obtained after the second stage of gel formation have a different
molecular arrangement from that of the original polymer.

VI. Conformation

Figure 12 shows some results of Ogawa and his colleagues.
They studied the conformational behavior of polysaccharide 13140
in alkaline solution by measuring the optical rotatory dispersion,

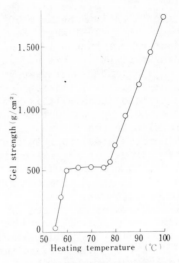

Agricultural and Biological Chemistry

Figure 5. Effect of heating temperature on gel strength of curdlan (22)

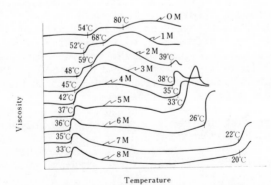

Temperature

Figure 6. Effects of urea (0–8 M) on gel formation of curdlan (1%). Shimazu microviscograph type SN1 was used.

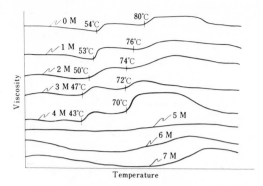

Figure 7. *Effects of ethylene-glycol (0–7 M) on gel formation of curdlan (1%). Shimazu microviscograph type SN1 was used.*

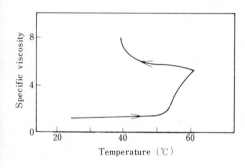

Figure 8. *Effect of heating temperature on specific viscosity of polysaccharide 13140 (1%)*

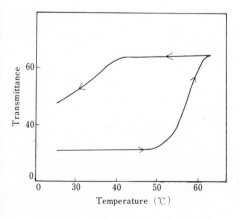

Figure 9. *Effect of heating temperature on transmittance of polysaccharide 13140 (1%)*

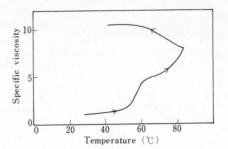

Journal of Food Science

Figure 10. Effect of heating temperature on specific viscosity of polysaccharide 13140 (1%) (24)

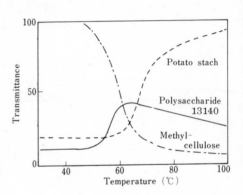

Figure 11. Effect of heating temperature on transmittance on polysaccharide 13140 (1%)

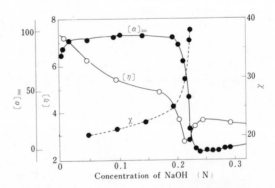

Carbohydrate Research

Figure 12. Dependence of optical rotation [α]₃₀₀ intrinsic viscosity [η] and extinction angle χ of a polysaccharide 13140 solution on concentration of sodium hydroxide. The value of χ is obtained in 0.5 g/100 ml glucan solution at the rate of shear of 6000 sec⁻¹ and 30°C (26).

vicosity and flow birefringence (26). All these characters were
found to change greatly in about 0.2 N sodium hydroxide. Thus,
they proposed that at low concentrations of sodium hydroxide, this
polymer has an ordered conformation, whereas at higher sodium
hydroxide concentrations, it consists of random coils and that the
transition of conformation occurs in about 0.2 N sodium hydroxide.
Saito and Sasaki confirmed this proposal, using ^{13}C NMR as shown
in Figure 13 (27).

Increase in the salt concentration causes a conformational
transition of the glucan from random coils to an ordered structure
when the concentration of alkali is below 0.3 N (28). Change in
conformation of curdlan in a solution of dimethyl-sulfoxide to a
rigid ordered structure occurred on addition of nonsolvents, such
as 2-chloroethanol, dioxane or water (29). Ogawa and his
colleagues also showed that the optical rotation depended upon the
degree of polymerization of the glucans in 0.1 M sodium hydroxide
(Figure 14) (30). The optical rotation of the soluble fractions
and that of the original glucans are practically constant, whereas
that of the insoluble fractions increased with DPn. Thus, they
concluded from their results that the glucan takes an ordered form
at low concentration of alkali when the DPn of the glucan is above
25. The content of the ordered form increases with DPn until it
reaches a maximum value and becomes constant at DPn values of
about 200. This may be the lower limit of DPn for gel formation
in neutral media. Formation of a complex of curdlan with Congo
Red was studied by Ogawa and his colleagues (31) and by Ogawa and
Hatano (32) by measuring the circular dichroism spectra in the
visible region. This results suggested that there may be two
kinds of binding systems in alkaline media.

The supramolecular structures of glucans with different
degrees of polymerization were compared by electron microscopy
using heated and unheated samples (Figure 15) (33). Microfibrils
of 100 - 200 Å width composed of many elementary fibrils, were
observed in the original and depolymerized glucan (DPn 400 and 260)
and the insoluble fraction of higher molecular weight (DPn 140),
but no microfibrils were detectable in the insoluble fraction of
low molecular weight (DPn 36) or the soluble fraction (DPn 13).
The microfibrils of the original glucan were very much longer than
those of the insoluble glucan. Thus, only glucans with higher
degrees of polymerization can form a gel when heated, form a
complex with Aniline Blue or Congo Red, show higher optical
rotations and form microfibrils seen by electron microscopy. No
significant difference was observed between heated and unheated
preparations. Jeisma and Kreger reported that β-1,3-glucan from
rhizomorph of Armillana mellea also showed an X-ray fiber pattern
and did not form a resilient gel on heat treatment (34).

Three types of water seem to be involved in the gel; one is
in the structure of elementary fibrils, another between the elementary fibrils and the third between the microfibrils. The
existence of three types of water in this gel was proposed by

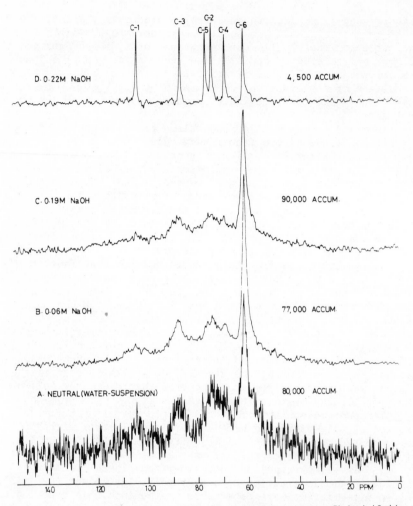

Abstract of the Annual Meeting of Japanese Biochemical Society

Figure 13. ^{13}C *NMR spectra in water suspension and alkaline solution (90°
pulse, repetition time 0.6 sec) of polysaccharide 13140 (27)*

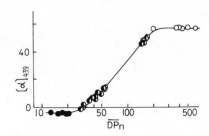

Carbohydrate Research

*Figure 14. Dependence of the spe-
cific rotation of polysaccharide 13140
at 439 nm on the degree of polymerization in 0.1M sodium hydroxide at
30°C. (○) gel-forming β-glucans;
(◐) insoluble fractions; (●) soluble
fractions (30).*

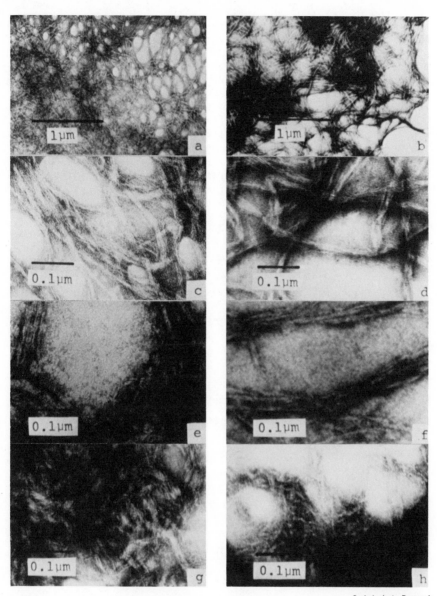

Carbohydrate Research

Figure 15. β-1,3-Glucan (polysaccharide 13140) microfibrils negatively stained with uranyl acetate. (a, b, c, d) original glucan ($\overline{DPn}$ 400); (d, e) depolymerized glucan ($\overline{DPn}$ 260); (f, g) depolymerized glucan ($\overline{DPn}$ 140). b, d, f, and h were heated at 95°C for 10 min; a, c, e, and g were not (33).

Suzuki and Aizawa from NMR analyses (35).
     The X-ray diffraction patterns of this polymer were analyzed
by Takeda and his colleagues (Figure 16) (36). The presence of
water in the preparation is required for X-ray analysis. It is
noteworthy that the center of the diffraction pattern has a cross-
like appearance, suggesting that the molecule has a rather simple
helical structure. Saito and Sasaki have succeeded to observe
broad carbon-13 resonance peaks with line-width, C.A. 150 H_z (C1
- C5) and H_z (C6) in gel of curdlan (27). The profils of
carbon-13 NMR in the gel of curdlan and in the solution of
degraded curdlan ($\overline{DPn}$ 13) were compared. It is interesting that
the carbon-13 peaks of C1 and C3 in gel, shifted downfield. These
shifts could be explained by a preferred rotamer population around
the β-1,3-glucosidic linkages. Their studies also supported the
presence of a single helix in addition to multiple helix(Fig. 17).

VII. Application

     The potential uses of curdlan in food products are shown in
Table 3.

Table 3
Application in Food Products

Function	Food
Food material	
Gelling agent	Jelly, jelly-like food, sherbet, custard and dry mixes
Slimming additive (non-caloric material)	Dietetic and diabetic foods
Film and fiber former	Edible films and fibers
Food additive	
Improvement of viscoelasticity	Spaghetti and noodles
Binding agent	Hamburgers and starchy jelly
Water-holding agent	Sausages, ham and starchy jelly
Masking malodors or aromas	Boiled rice
Retention of shape	Starchy jelly and dry dessert mixes
Thickener and stabilizer	Salad dressings and spreads

     The polymer seems to be useful for gelling materials, for
jelly products and as a food additive for improving the quality of

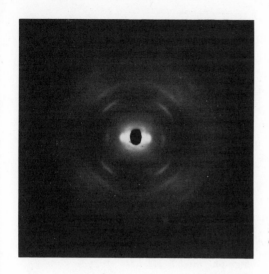

*Figure 16. X-ray diffraction analysis
of polysaccharide 13140 (wide-angle
x-ray diffraction patterns) (36)*

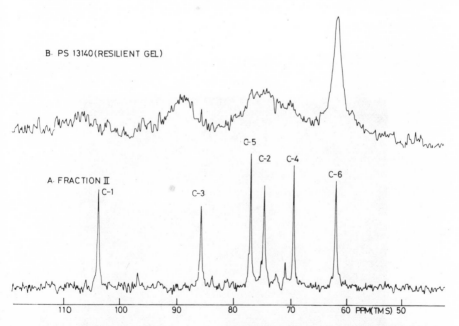

*Figure 17. ^{13}C NMR spectra in gel of polysaccharide 13140 and in solution of the de-
graded polymer ($\overline{DPn}$ 13) (27)*

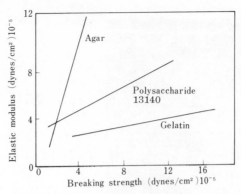

*Figure 18. Relationship between the break-
ing strength and elastic modulus of various
gels. Samples of gels were sliced into cylindri-
cally shaped pieces 23 mm thick and then
measured with Autograph model IM-100
(Shimazu)(24).*

Figure 19.   Gel of polysaccharide 13140

various foods. The polymer can be added during the production process before heating either as a powder, or as a suspension or slurry in water or aqueous alcohol. Various kinds of tests showed that this polymer is safe. The gel of curdlan has properties intermediate between the brittlenes of agar gel and the elasticity of gelatin (Figure 18) (24). This gel was shown to adsorb tannin (38).

In addition, nutrition studies on this polymer showed that it has no calory value. Thus it is useful as an ingredient of low-caloric foods. For example, a gel such as shown in Figure 19 can be easily made using this polymer.

It seems to have many potential industrial uses as a film, fiber or support for immobilized enzymes, as seen in Table 4.

Table 4
Application in Industry

Application	Function	Characteristic
Film or fiber	Material	Transparent, edible, insoluble in hot water and impermeable to oxygen
Tobacco products	Binding agent	Combustible and insoluble
Artificial feeds for fishes and silk worms	Gelling agent	Good texture
Molecular sieve and support for immobilizing enzymes	Carrier	

Films and fibers of this polymer are easily prepared and have the characteristics of being edible, insoluble in water, biodegradable and impermeable to oxygen. Beads and fibers of the polymer are good supports for immobilized enzymes (39, 40).

Extensive studies on various aspects of polysaccharide production are required. These may clarify the role of polysaccharides in nature and so indicate the values of these compounds for human life. Extensive studies have been made in many laboratories in Japan on biological, organic and physical chemical and microbiological aspects of curdlan and also on its use in food products and medicines and in industrial chemistry.

Literature Cited

1. Harada, T. and Yoshimura, T. J. Ferment. Technol. (1964) 42, 615.

2. Harada, T. and Yoshimura, T. Biochim. Biophys. Acta. (1964)
    83, 374.
3. Harada, T., Yoshimura, T., Hidaka, H. and Koreeda, A. Agr.
    Biol. Chem. (1965) 29, 757.
4. Harada, T. Arch. Biochem. Biophys. (1965) 112, 65.
5. Harada, T. and Yoshimura, T. Agr. Biol. Chem. (1965) 29, 1027.
6. Misaki, A., Saito, H., Ito, T. and Harada, T. Biochemistry
    (1969) 8, 4645.
7. Saito, H., Misaki, A. and Harada, T. Agr. Biol. Chem. (1970)
    34, 1683.
8. Harada, T., Masada, M., Hidaka, H. and Takada, M. J. Ferment.
    Technol.(1966) 44, 20.
9. Harada, T., Masada, M., Fujimori, K. and Maeda, I. Agr. Biol.
    Chem. (1966) 30, 196.
10. Harada, T., Misaki, A. and Saito, H. Arch. Biochem. Biophys.
    (1968) 124, 292.
11. Hisamatsu, M., Amemura, A., Harada, T., Nakanishi, I. and
    Kimura, K. Abstract of the Annual Meeting of Agr. Chem. Soc.
    Japan (1976) p289.
12. Nakanishi, I., Kanamaru, T., Kimura, K., Matsukura, A., Asai,
    M., Suzuki, T. and Yamotodani, S. 284th Meeting of Kansai
    Branch of Agr. Chem. Soc. Japan, Osaka (1972).
13. Nakanishi, I., Kimura, K., Suzuki, T., Ishikawa, M., Banno,
    I., Sakane, T. and Harada, T. J. Gen. Appl. Microbiol. (1976)
    22, 1.
14. Nakanishi, I., Kimura, K., Kusui, S. and Yamazaki, E. Carbohyd.
    Res. (1974) 32, 47.
15. Saito, H., Misaki, A. and Harada, T. Agr. Biol. Chem. (1968)
    32, 1261.
16. Ebata, J. Abstract of 8th International Symposium on Carbohyd.
    Chem., Kyoto (1976) p112.
17. Aspinall, G. O. and Kessler, G. Chem. Ind. (London) (1957)
    1296.
18. Harada, T., Fujimori, K., Hirose, S. and Masada, M. Agr. Biol.
    Chem. (1966) 30, 764.
19. Harada, T., Fujimori, K. and Masada, M. J. Ferment. Technol.
    (1967) 45, 145.
20. Amemura, A., Moori, K. and Harada, T. Biochim. Biophys. Acta.
    (1974) 334, 398.
21. Harada, T., Yamauchi, H., Hisamatsu, M., Ott, I., Nakanishi,
    I. and Kimura, K. Abstract of the Annual Meeting of Agr. Chem.
    Soc. Japan (1976) p288.
22. Maeda, I., Saito, H., Masada, M., Misaki, A. and Harada, T.
    Agr. Biol. Chem. (1967) 31, 1184.
23. Aizawa, M., Takahashi, M. and Suzuki, S. Chem. Letters (1974)
    193.
24. Kimura, H., Moritaka, S. and Misaki, M. J. Food Sciences
    (1973) 38, 668.
25. Konno, A., Azeti, Y. and Kimura, H. Abstract of the Annual
    Meeting of Agr. Chem. Soc. Japan (1974) p310.

26. Ogawa, K., Watanabe, T., Tsurugi, J. and Ono, S. Carbohyd. Res. (1972) 23, 399.
27. Saito, H. and Sasaki, T. Abstact of the Annual Meeting of Japanese Biochemical Soc.,(1976) p651.
28. Ogawa, K., Tsurugi, J. and Watanabe, T. Chem. Letters (1973) 95.
29. Ogawa, K., Miyagi, M., Fukumoto, T. and Watanabe, T. Chem. Letters (1973) 943.
30. Ogawa, K., Tsurugi, J. and Watanabe, T. Carbohyd. Res. (1973) 29, 397.
31. Ogawa, K., Tsurugi, J. and Watanabe, T. Chem. Letters (1972) 689.
32. Ogawa, K. and Hatano, M. 288th Meeting of Kansai Branch of Agr. Chem. Soc. Japan, Osaka (1974).
33. Koreeda, A., Harada, T.., Ogawa, K., Sato, S. and Kasai, N. Carbohyd. Res. (1974) 33, 396.
34. Jeisma, J. and Kreger, D. R. Carbohyd. Res. (1975) 43, 200.
35. Suzuki, S. and Aizawa, M. Abstract of 8th International Symposium on Carbohyd. Chem., Kyoto (1976) p76.
36. Takeda, H., Yasuoka, W. Kasai, N. and Harada T. 283rd Meeting of Kansai Branch of Agr. Chem. Soc. Japan (1973).
37. Nakagawa, T., Moritaka, S. and Kimura, H. Abstract of the Annual Meeting of Agr. Chem. Soc. Japan (1973) p196.
38. Nakabayashi, T. Nihon Shokuhin Kogyo Kaishi (1974) 21, 341.
39. Takahashi, K., Yamazaki, Y., Kato, K and Takahashi, T. Abstract of the Annual Meeting of Agr. Chem. Soc. Japan (1976) p401.
40. Murooka, Y., Yamada, T. and Harada, T. Annual Meeting of Ferment. Technol. Japan, Osaka (1976).

21

Dextrans and Pullulans: Industrially Significant α-D-Glucans

ALLENE JEANES

Northern Regional Research Center, Agricultural Research Service,
U.S. Department of Agriculture, Peoria, IL 61604

This symposium on extracellular microbial polysaccharides of practical importance would not be complete without consideration of the α-D-glucans, dextran and pullulan.

The significance of dextran in man's practical affairs was apparent before its origin, identity, and name were established. Dextrans develop naturally in sucrose-containing solutions that have become inoculated with dextran-producing bacteria from air, plants, or soil. The resulting transformation of the solutions to syrupy, viscous, or ropey fluids, or even to gelled masses, doubtlessly has plagued man since the inception of accumulating and storing sucrose-containing foods and beverages. As early as 1813 (1,*2*)[1] , reports described the mysterious thickening or solidification of cane and beet sugar juices, and later impediment of filtration and crystallization was traced to the occurrence of this condition. In 1861, Louis Pasteur (3) initiated systematic scientific progress by explaining that these "viscous fermentations" resulted from microbial action. In 1878, van Tieghem (4) named the causative bacteria Leuconostoc mesenteroides because its growth in colorless flocs resembled that of the green algae of the genus Nostoc. In 1880 (5), Scheibler established this type of product as a glucan having positive optical rotation and named it dextran.

Thus, through the importance of the dextran class of α-D-glucans in man's economy, the dextrans were the first extracellular microbial polysaccharides to come under systematic scientific investigation. Dextrans from several bacterial strains also were the first extracellular microbial polysaccharides to be produced and used industrially. A comprehensive review (6) was published in 1966 on dextran production,

---

[1]References other than reviews, which cite original research publications not included here, are marked with an asterisk.

structure, properties, uses, and related considerations.
The same topics were reviewed from a different viewpoint in
1973 (7). The biosynthesis and structure of dextrans was
reviewed comprehensively in 1974 as well as the structurally
dependent specific interactions with immunoglobulins (antibodies)
and globulins such as concanavalin A (8). An extensive
bibliography on all scientific aspects of dextran (exclusive
of clinical research and testing) and dextran derivatives
includes information from 1861 through mid-1976 (9).

Summarized here is the current status of dextran as an
established product of world commerce and in relation to
specific industries.

Interest in pullulan and its practical potentialities
have developed since 1959 when Bender, Lehman, and Wallenfels
(10) first characterized the water-soluble extracellular
product from Aureobasidium (Pullularia) pullulans and named
it accordingly. The polysaccharide had been isolated
previously and partially characterized in studies of micro-
organisms responsible for breakdown of forest litter (11).
The slime-forming black yeastlike fungus, A. pullulans,
occurs ubiquitously in organic waste matter which it decomposes
in soil, rivers, paper-mill effluents, and sewage (12). The
microorganism has adverse economic importance because of its
costly deterioration of paint, discoloration of lumber, and
attack on plants and plant products (13). In none of these
natural occurrences, however, does the polysaccharide seem
to have a role except as a slimy nuisance. Already of
applied practically, however, is the enzyme pullulanase
(pullulan 6-glucanohydrolase EC 3.2.1.41) which was discovered
by Bender and Wallenfels in Aerobacter aerogenes (14) and
shown to depolymerize pullulan to its repeating unit maltotriose
by specific attack on the interunit α-1,6-linkages. Pullulanase,
now obtainable in practical amounts from numerous microbial
sources, also cleaves α-1,6-linkages in starch and is used
industrially to release the unit chains in starch (15,16).
Substrates other than pullulan, however, may be used for
producing pullulanase.

The production, properties, and potential uses of
pullulan have been reviewed (12). Summarized here are the
constitutional bases for practical applications and the uses
that have been proposed.

Importance of Naturally Occurring Dextrans

The predilection of Leuconostocs for sucrose in nature
has a specific basis. Sucrose induces in these bacteria
formation of the dextran-synthesizing enzyme dextransucrase
(sucrose: 1,6-α-D-glucan 6-α-glucosyltransferase, E.C.
2.4.1.5). This enzyme accomplishes dextran synthesis by

transferase action without need for intermediate substrates.
Fructose, the byproduct of dextran synthesis, is metabolized
by Leuconostocs which cannot, however, metabolize either
sucrose (they have neither invertase nor sucrose phosphorylase)
or dextran. Extracellular dextransucrase is produced
abundantly by many strains, although the dextransucrase of
some strains is cell-bound. The rate and extent of activity
on sucrose that may result is illustrated dramatically by an
historic report of the fortuitous conversion of 5000 liters
of molasses to a compact gel mass in 12 hours (4).

In 1972, the status of the situation was that, "although
it is difficult to quantify the effects of polysaccharides
on the economics of sugar cane processing, it is obvious
from the volume of recent literature that importance is
attached to their elimination from the process" (17). The
dextrans have a major role (9,17) although bacterial levans
and polysaccharides of plant origin are involved also (17).
The long-known adverse effects of dextran continue in
polarization measurements, clarification and filtration, and
in reducing the rate and efficiency of crystallization.
In addition, traces of dextran cause inferior crystal structure
by elongating the c axis (18,19). The beet sugar industry
is less affected by dextran contamination. Sucrose is less
exposed to infection during harvesting and first stages of
processing of beets than of cane.

Very sensitive biochemical tests have demonstrated the
extent of dextran contamination in commercial sucrose,
including that distributed as a standard of highest purity.
Neill, Hehre, and coworkers (20) demonstrated serological
activity indicative of dextran in both cane and beet sugars
from diverse geographical sources and various methods of
manufacture. The majority of the cane products showed
higher serological activity than did the majority of the
beet products. The weakest activities were in several
samples of reagent grade sucrose of German origin prepared
from beet sugar. Gibbons and Fitzgerald (21), utilizing the
agglutinizing action of dextran on cells of Streptococcus
mutans, also demonstrated dextran in reagent-grade sucrose.

Dextranases are being investigated (22) and used (23)
for removal of dextran from cane sugar juices as well as
from sucrose solutions and wines made hazy by the presence
of dextran.

Constitutional Basis for Practical Importance

Dextran. By definition, the generic name dextran
applies to a large class of α-D-glucans in which predominance
of α-1,6-linkages is the common feature. One of the simplest
dextrans known is that from Leuconostoc mesenteroides NRRL
B-512(F); the structural features are shown in Figure 1.

The α-$\underline{\text{D}}$-glucopyranosidic linkages are 95% 1,6-and 5% 1,3-
(24). The 1,3-linkages are points of attachment of side
chains of which about 85% are 1 or 2 glucose residues in
length (25). The remaining 15% of the side chains may have
an average length of 33% glucose residues and may not be
uniformly distributed in the macromolecule (26). This
dextran is readily soluble in water; certain other dextrans
may be insoluble. In dextrans from other strains, the non-
1,6-linkages may be 1,2-, 1,3-, or 1,4-. Only one type may
occur in a dextran, or there may be two or three. Great
diversity is thus created. Dextran available in the United
States and western Europe is produced from sucrose by strain
NRRL B-512(F). Dextrans of apparently similar structure,
but from different strains, are produced in Japan (27) and
Russia (28). Dextrans produced in other countries of Europe
and Asia are from selected strains (29*).

Dextran having this structure (Figure 1) was selected
for production because the fraction ($\overline{M}_w$ 75,000 + 25,000)
prepared from it for intravenous administration (blood
volume expander) was substantially less antigenic as compared
with that from dextrans having higher percentages of non-
1,6-linkages. Dextran from strain NRRL B-512(F) is completely
metabolized in man (30) when either ingested or administered
parenterally as a fraction of suitable molecular size and
size distribution. Derivatization, however, slows or inhibits
metabolism.

An additional asset of strain NRRL B-512(F) is its
copious formation of dextransucrase (31). Production of
dextran by use of cell-free culture filtrates rather than in
growing cultures results in enhanced yield, quality and ease
of purification of the product. And furthermore, by suitable
adjustment of conditions, the major product can be synthesized
directly within a chosen molecular weight range (32, 9).

The native dextran may have weight-average molecular
weight ($\overline{M}_w$) values (33) (light scattering) of 35-50 X 10^6.
The structural simplicity of this dextran permits graded
partial depolymerization and separation into fractions of
any desired $\overline{M}_w$ and size distribution, which differ primarily
in molecular weight. The content of branch points remaining,
however, would depend on the method of partial depolymerization;
it is decreased by acid hydrolysis (34) but not by use of
endo-acting dextranases (1,6-α-$\underline{\text{D}}$-glucan 6-glucanohydrolase,
E.C. 3.2.1.11). Fractions of lower molecular weight obtained
through such enzymolysis retain the branch points, and their
structural details would be determined by the action pattern
of the specific dextranase (35). The series of fractions
produced from partial depolymerizates is unique. Selected
fractions or derivatives of them serve pharmaceutical or
other purposes having specific requirements for molecular

size in order to achieve physiological compatibility or
other special objectives.

Production of such fractions by direct, controlled
enzymatic synthesis is not known to be in use.

The high proportion of 1,6-linkages in dextran NRRL B-
512(F) confers unusual flexibility on the chain and leaves
numerous sites for substitution, essentially all of which
are in secondary positions. The ratio of relative rate
constants established for methylation of the hydroxyl groups,
$C_2:C_3:C_4:8:1:3.5$ (36) indicates also the relative reactivity
towards other substituents such as the sulfate (37). The
frequent occurrence of three hydroxyl groups in consecutive
positions in the glucopyranosidic residues of dextrans would
appear to account for their unusual ability to complex with
large amounts of metallic elements such as ferric iron and
calcium. Such complexes are important as pharmaceutical
preparations and in certain metallurgical processes.

Thus, the charcteristics of dextran from strain NRRL B-
512(F) that determine its value in practical applications,
reside in its composition as a soluble α-D-glucan and in the
properties of its primary structure. In contrast, it has
been emphasized in this symposium that the unique characteristics
of the anionic heteropolysaccharide xanthan which are basic
to its usefulness, result from secondary and tertiary structural
effects (38,39,40). The specific role of ionic charge,
which also may be influential in xanthan properties, has not
been established but may be inferred from research on ionogenic
derivatives of dextran (41).

Pullulan. The generic name pullulan is applied to any
extracellular α-D-glucan elaborated by A. pullulans from a
variety of substrates. A commonly observed feature is the
predominant repeat unit maltotriose polymerized linearly
through 1,6-linkages (Figure 2). Frequently present also
are α-maltotetraose units (42,43,44) contained mainly
within the polymer chain (43) in amounts of 6.6% (43) and
5-7% (44). In products in which possible heterogeneity was
not excluded, traces of other neutral sugars and uronic
acids have been reported (12). Products from other strains
and from other genera and species have shown variation on
the basic pullulan pattern such as the presence of 1,3-
linked glucosyl residues (45,46). Thus, "there is, perhaps,
no unique structure of pullulan" (43).

The molecular weight of a pullulan product differs with
the length of fermentation time (47,48). Molecular weight
of 2×10^6 developed initially during limited fermentation
decreased to 1.5×10^5 during continued fermentation (47).
The site of degradation is the internally located maltotetraose
units; the degradative enzyme appears to be an "endoamylase"
produced during culture growth (43,47). The modified

pullulan resulting from "endoamylase" action is inert to α-amylase (43).

Such uncontrolled variation in molecular weight can be eliminated, however, by choice of strain and adjustment of the pH and of the phosphate content of the culture medium (49). Pullulan products having molecular weights as high as 250×10^4 or as low as 5×10^4 may be obtained in this way.

The mechanism of biosynthesis of pullulan discourages consideration of enzymatic synthesis as a means for production. Synthesis is accomplished through mediation of sugar nucleotide/lipoid carrier intermediates associated with cell membrane fractions (50).

The pullulan molecule may be considered as a chain of amylose, the linear component of starch, in which an α-1,6-bond replaces every third α-1,4-bond. The 1,6-bond introduces flexibility, and the interrruption of regularity results in making pullulan readily soluble, eliminating retrogradation and improving rather than impairing fiber- and film-forming ability. The presence of the 1,6-bonds may influence the position of substituents and properties of derivatives by introducing a different sequence of free hydroxyl groups. The presence of 1,6-linkages, spaced as they are, prevents attack by salivary and intestinal amylases (43). Isoamylase from Pseudomonas sp., which cleaves α-1,6-bonds in amylopectin and glycogen, also is inert on pullulan (51).

Dextran and Dextran Derivatives in Industry

Pharmaceutical Industry. Probably the largest outlet for dextran and dextran derivatives is through the pharmaceutical and fine chemicals industries. The major developments have originated from fundamental research in Sweden which was initiated about 1944 and has continued consistently (52,53). Research and development have followed, however, in numerous other countries throughout the world which produce their own pharmaceutical products from dextran (9,27,28,29*).

Two dextran fractions of major significance are used in suitably prepared solutions for parenteral administration (6,9). The fraction of $\overline{M}_w$ 70,000 is used to restore and maintain blood volume in treatment of shock, hemorrhage, extensive burns, and a variety of other physiological conditions. The fraction of $\overline{M}_w$ 40,000 is used to improve flow in capillaries, treatment of vascular occlusion, artificial extracorporeal perfusion of organs, and in a variety of other ways.

These and other sharply cut dextran fractions are used for preparation of numerous derivatives such as the sulfates, diethylaminoethyl (DEAE) dextran, and complexes with iron and other metallic elements. These substances serve a variety of purposes (9,54,55). Dextran sulfates have anticoagulant, antilipemic, and antiulcer activity. They are used in liquid two-phase separation and concentration of living cells such as those of viruses, blood, tumors, and other tissues. DEAE dextran enhances biological effects of macromolecules and vaccines. A soluble complex of dextran and iron is produced widely in numerous countries for intra-muscular administration to alleviate iron-deficiency anemia in the human and in domestic animals. The solution contains 5% iron and 20% dextran of $\overline{M}_w$ 5,000 (56). The iron is mainly nonionic (56) and appears to be β-FeOOH (57). The initial patents (58) have been emulated extensively (9). A soluble calcium complex containing 10-12% calcium is administered parenterally to alleviate hypocalcemia of cattle delivery paresis (9). Complexes with antimony and arsenic are effective against tropical infections (9).

Crosslinked dextran gels or their ionic derivatives are employed in purification, fractionation and isolation of enzymes, hormones, and other sensitive biological substances without modification of their activity. By covalent bonding to either dextran or crosslinked dextran gels, enzymes, immunoglobulins, and antigens are stabilized and supported for use in specific reactions (59,9).

Dextranases, prepared by growth of various molds on dextrans, are used in mouthwashes and toothpaste to either disperse or inhibit formation of dento-bacterial plaques which contain dextrans and foster carious dental lesions (9,60,61,62).

Food Industry. The potentialities for dextran in the food industry have been reviewed (63,64). The only actual uses known to the author, however, are in dextran gel-filtration processes to concentrate proteins or to recover proteins from liquid wastes and effluent streams. From cereal waste streams, 70% recovery of protein has been effected. In the milk industry, skim milk or cheese whey is fractionated for recovery of undenatured protein components of enhanced quality, nutritive value, and applicability. A plant having capacity of 1 X 10^6 lb. per day is in operation (65). Protein is separated from lactose and mineral constituents and fractionated mainly on the basis of molecular weight into casein, β-lactoglobulin, and α-lactoglobulin. β-Lactoglobulin, which is highly superior nutritionally to casein (66), had restricted use when previously isolated as the degraded and denatured lactalbumin (67). Other valuable products that may be recovered are lactoferrin and immuno-globulins.

By another application of gel filtration, the protein content of milk is increased from 3.35% to 5.35% without increase of the lactose and mineral contents (68).

Atomic Fuel and Metallurgy. Gel precipitation is a process in which dextran (or certain other polyhydric polymers) is used to produce a metal compound in the form of a gel under conditions where an insoluble precipitate would be expected (69) (Figure 3). The process is used for purifying, separating, and concentrating metals from solutions of their salts or mixtures of salts or from colloidal dispersions of aqueous hydrous sols. The final product may be in the form of powder, granules, spheres, rods or shaped rods, or ceramic coatings and moulded objects. Products prepared by use of dextran as the gelling agent are for use as nuclear reactor fuels (70,71), catalysts (71,72,73), ceramic coatings (70), refractories and ferro-electric materials (72,73), and powder for alloys (72), pigments (74) and metallurgical processes (70).

The gelling agent and metal ions appear to form molecular complexes which, when contacted with the precipitating reagent, produce discrete microcrystalline gel particles (69,75). The 1,6-linkages in dextran are believed to confer a special configuration on the three contiguous free -OH groups which is peculiarly favorable to -OH--complex formation with an unusual number of metal ions (69,76).

The specific properties of dextran metallic ion complexes are utilized in separating ferric iron from mixtures with copper, nickel or cobalt, or nickel from thorium, or zirconium from copper (76). Dextran (or fractions of stated molecular weight) is utilized in preparing black magnetic iron oxide (Fe_3O_4) from ferrous salt (74,77). Cupric ion may be adsorbed from solution on hydrous gels of ferric oxide, chromium oxide, or thorium phosphate/dextran gel, and then eluted (77).

The procedures reviewed here indicate the potential for dextran application in gel-precipitation processes. Some of the procedures are known to be in use.

Petroleum Production. A pioneering concept advocated for some years was to make dextran a profitable byproduct of the sugar cane industry by using it in petroleum drilling muds (78). In initial laboratory testing for water loss inhibition, a modified dextran gave results equivalent or superior to starch and carboxymethyl cellulose (79). The modified dextran (Viscoba), prepared by treatment of dextran with aldehyde before isolation (80), had improved viscosity and was resistant to microbial attack. The concept was advanced further when, during 1956 through 1959, a dextran production pilot plant was operated in conjunction with a sugar mill in Cuba (81). The dextran, produced from strain NRRL B-512(F) by a modified enzymatic procedure, was precipitated once and drum dried. [The fructose byproduct was recovered and uses investigated (81)]. The output (3-6 tons/day) was used in the United States in drilling muds under a variety of field conditions. The price at the well site was 46 cents/lb.; the demand greatly exceeded the supply (82). The dextran

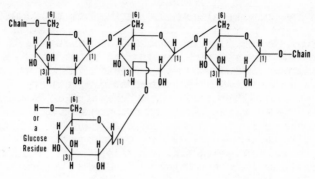

Figure 1. Structural features of dextran from Leuconostoc mesenteroides NRRL B-512(F)

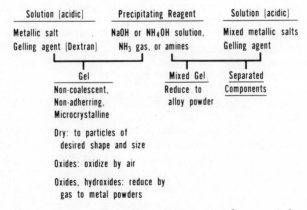

Figure 2. The characteristic structural features of pullulans: α-maltotriose polymerized through α-1,6-linkages

Solution (acidic)	Precipitating Reagent	Solution (acidic)
Metallic salt	NaOH or NH₄OH solution,	Mixed metallic salts
Gelling agent (Dextran)	NH₃ gas, or amines	Gelling agent

Gel	Mixed Gel	Separated
Non-coalescent,	Reduce to	Components
Non-adherring,	alloy powder	
Microcrystalline		

Dry: to particles of
   desired shape and size

Oxides: oxidize by air

Oxides, hydroxides: reduce by
   gas to metal powders

Figure 3. The gel-precipitation process and some of the products resulting

functioned better than some of its competitors and as well
as any.[2/]

The dextran could not be used in lime base muds; it
precipitated at pH 11.0-11.2 and lost its water-binding
capacity. Under these conditions, however, lightly hydroxy-
ethylated dextran retained its water-binding capacity (82).
At less basic or neutral pH, dextran tolerates calcium ion
and magnesium ion well (83). Dextran is the hydrocolloid in
an "inhibited mud" composition containing 3500 ppm calcium
ion that is used to inhibit shale hydration (84). Like all
polymers, dextran is susceptible to free radical degradation,
and protection is advised during processing as well as use
(85).

Dextran NRRL B-512(F) has properties suitable for use
in viscous water flooding (86); in screening tests, it gave
results superior to many other substances examined. The
unfavorable results of a field trial (87) may have related
to lack of protection against free-radical degradation.

Photographic Industry. Native high molecular weight
dextran has been supplied consistently for an undisclosed
industrial use believed to relate to photographic products.
Numerous patents have been issued in the past and continue
to be issued on the superior effects achieved from certain
dextran derivatives in X-ray and photographic emulsions
(9,54,55). It seems probable that some of these derivatives
are in use.

Pullulan--Proposed Uses

Numerous applications of pullulan and its derivatives
have been proposed and patented, but apparently are not yet
in use (88,89). The possibility of eventual success for
most uses is increased by the claim that, by proper selection
of pullulan-producing strains, the molecular weight can be
controlled and absence of black pigment in the product can
be assured (49).

The efficiency of pullulan, even as the crude fermentation
liquors, has been demonstrated for flocculation of clay
slimes from aqueous solutions resulting from beneficiation
of uranium, potash, and other ores (90,91,92).

Films formed from pullulan without plasticizers have
excellent physical properties, are water-soluble, impervious
to oxygen and suitable for coating or packaging foods and
pharmaceuticals especially when exclusion of oxygen is
desirable (88). Fibers from pullulan have a shiny gloss and
high tensile strength which, after stretching, is described

[2/] Death of the key personnel in an airplane accident
terminated this development.

as comparable to that of nylon (88). The fibers may be admixed with natural fibers in special papers and other products (93). Pullulan is suitable for making adhesives and shaped articles by compression molding. In such molded articles pullulan has characteristics similar to polyvinyl alcohol or to styrene (88). It has desirable properties for use in noncaloric and other foods; it is nontoxic and non-digestible (88). Pullulan is biodegradable, however, under usual conditions of waste disposal.

Literature Cited

1. Browne, C. A., Jr., J. Am. Chem. Soc. (1906) 28, 453-469.
2. Hehre, E. J. and Neill, J. M., J. Exp. Med. (1946) 83, 147-162.
3. Pasteur, L., Bull. Soc. Chim. Paris (1861), 30-31.
4. van Tieghem, P., Ann. Sci. Nat. Bot. Biol. Veg. (1878) 7, 180-203.
5. Scheibler, C., Ver. Rubenzucker-Ind. (1874) 24, 309-335.
6. Jeanes, A., "Dextran," in Encyclopedia of Polymer Science and Technology, Vol. 4, Bikales, N. M., Ed. Interscience Publishers, N.Y., 1968, pp. 693-711.
7. Murphy, P. T. and Whistler, R. L., "Dextrans," in Industrial Gums: Polysaccharides and their Derivatives, Second Edition, Whistler, R. L. and BeMiller, J. N., Eds. Academic Press, N.Y., 1973, pp. 513-542.
8. Sidebotham, R. L., Adv. Carbohydr. Chem. Biochem. (1974) 30, 371-444.
9. Jeanes, A., "Dextran Bibliography: Extensive Coverage of Research Literature (Exclusive of Clinical) and Patents, 1861-1976." Miscellaneous Publication, Agricultural Research Service, United States Department of Agriculture, in press.
10. Bender, H., Lehmann, J., and Wallenfels, K., Biochim. Biophys. Acta (1959) 36, 309-316.
11. Bernier, B., Can. J. Microbiol. (1958) 4, 195-204.
12. Zajic, J. E. and LeDuy, A., "Pullulan," in Encyclopedia of Polymer Science and Technology, Supplement Vol. 2, Bikales, N. M., Ed. Interscience Publishers, N.Y., in press, 1977.
13. Cook, W. B., Mycopathol. Mycol. Appl. (1959) 12, 1-45.
14. Bender, H. and Wallenfels, K., Biochem Z. (1961) 334, 79-95.
15. Enevoldsen, B. S., J. Inst. Brew., London (1970) 76, 546-552; Brygmesteren (1971) 28, 41-51.
16. Hathaway, R. J. (A. E. Staley Mfg. Co.), U.S. Patent 3,556,942. January 19, 1971.

17. Imrie, F. K. E. and Tilbury, R. H., Sugar Technol. Rev. (1972) 1, 291-361.
18. Sutherland, D. N. and Paton, N., Int. Sugar J. (1969) 71, 131-135.
19. Leonard, G. J. and Richards, G. N., Int. Sugar J. (1969) 71, 263-267.
20. Neill, J. M., Sugg, J. Y., Hehre, E. J., and Jaffe, E., J. Exp. Med. (1939) 70, 427-442; Am. J. Hyg. (1941) 34, 65-78.
21. Gibbons, R. J. and Fitzgerald, R. J., J. Bacteriol. (1969) 98, 341-346.
22. Richards, G. N. and Streamer, M., Carbohydr. Res. (1972) 25, 323-332.
23. Tate and Lyle Ltd., British Patent 1,290,694, September 27, 1972.
24. Van Cleve, J. W., Schaefer, W. C., and Rist, C. E., J. Am. Chem. Soc. (1956) 78, 4435-4438.
25. Larm, O., Lindberg, B., and Svensson, S., Carbohydr. Res. (1971) 20, 39-48.
26. Walker, G. J. and Pulkownik, A., Carbohydr. Res. (1973) 29, 1-14.
27. Misaki, A., Yukawa, S., Asano, T., and Isono, M., Ann. Rep. Takeda Res. Lab. (1966) 25, 42-54; Chem. Abstr. (1967) 66, 54,255t.
28. Rosenfel'd, E. L., Biokhimiya (1958) 23, 635-638; Biochem. English Transl. (1958) 23, 597-600.
29. Ewald, R. A. and Crosby, W. H., Transfusion (1963) 3, 376-386.
30. Jeanes, A., ACS Symp. Ser. (1975) 22, 336-347.
31. Koepsell, H. J. and Tsuchiya, H. M., J. Bacteriol. (1952) 63, 293-295.
32. Tsuchiya, H. M., Hellman, N. N., Koepsell, H. J. and others, J. Am. Chem. Soc. (1955) 77, 2412-2419.
33. Senti, F. R., Hellman, N. N., Ludwig, N. H., and others, J. Polym. Sci. (1955) 17, 527-546.
34. Lindberg, B. and Svensson, S., Acta Chem. Scand. (1968) 22, 1907-1912.
35. Walker, G. J. and Pulkownik, A., Carbohydr. Res. (1974) 36, 53-66.
36. Norrman, B., Acta Chem. Scand. (1968) 22, 1381-1385.
37. Miyaji, H. and Misaki, A., J. Biochem. (1973) 74, 1131-1139.
38. Dea, I. D. M. and Morris, E. R., this symposium.
39. Morris, E. R., this symposium.
40. Moorhouse, R., Walkinshaw, M. D., and Arnott, S., this symposium.
41. Pasika, W. M., this symposium.
42. Wallenfels, K., Keilich, G., Bechtler, G., and Freudenberger, D., Biochem. Z. (1965) 341, 433-450.

43. Catley, B. J. and Whelan, W. J., Arch. Biochem. Biophys. (1971) 143, 138-142.
44. Taguchi, R., Kikuchi, Y., Sakano, Y., and Kobayashi, T., Agric. Biol. Chem. (1973) 37, 1583-1588.
45. Sowa, W., Blackwood, A. C., and Adams, G. A., Can. J. Chem. (1963) 41, 2314-2319.
46. Elinov, N. P. and Matveeva, A. K., Biokhimiya (1972) 37, 255-257; Biochem. English Transl. (1973) 37, (2, Part 1), 207-209.
47. Catley, B. J., FEBS Lett. (1972) 20, 174-176.
48. LeDuy, A., Marsan, A. A., and Coupal, B., Biotechnol. Bioeng. (1974) 16, 61-76.
49. Kato, K. and Shiosaka, M. [Hayashibara Biochemical Laboratories, Inc.] U.S. Patent 3,912,591. October 14, 1975.
50. Taguchi, R., Sakano, Y., Kikuchi, Y., and others, Agric. Biol. Chem. (1973) 37, 1635-1641.
51. Yokobayashi, K., Akai, H., Harada, T. and others, Biochim. Biophys. Acta (1973) 293, 197-202.
52. Groenwall, A. and Ingelman, B., Acta Physiol. Scand. (1945) 9(1), 1-27.
53. Tiselius, A., Porath, J., and Albertsson, P. A., Science (1963) 141, 13-20.
54. Jeanes, A., J. Polym. Sci.: Polym. Symp. No. 45, Ion-Containing Polymers (1974), 209-227.
55. Jeanes, A. in "Polyelectrolytes," Frisch, K. and Klempner, D., Eds. Technomic Publishing Co., Inc., Wesport, Conn., 1977.
56. Cox, J. S. G., King, R. E., and Reynolds, G. F., Nature (London) (1965) 207, 1202-1203.
57. Marshall, P. R. and Rutherford, D., J. Colloid Interface Sci., (1971) 37, 390-402.
58. London, E. and Twigg, G. D. (Benger Laboratories, Ltd.), British Patent. 748,024, April 18, 1956; U.S. Patent 2,820,740, January 21, 1958.
59. Kagedal, L. and Akerstroem, S., Acta Chem. Scand. (1971) 25, 1855-1859.
60. Keyes, P. H., Hicks, M. A., Goldman, B. M., and others, J. Am. Dent. Assoc. (1971) 82, 136-141.
61. Miller, G. R. (Colgate-Palmolive Co.), U.S. Patent 3,630,924, December 28, 1971.
62. Woodruff, H. B. and Stoudt, T. H. (Merck and Co., Inc.), U.S. Patent 3,686,393, August 22, 1972.
63. Glicksman, M., "Gum Technology in the Food Industry," Academic Press, New York, 1969, pp. 335-341.
64. Jeanes, A., Food Technol. (1974) 28(5), 34-40.
65. Davis, J. C., Chem. Eng. (July 1972), 114-115.
66. Forsum, E., J. Dairy Sci. (1973) 57, 665-670.
67. Wingerd, W. H., J. Dairy Sci. (1971) 54, 1234-1236.

68. Samuelsson, E-G., Tibbling, P., and Holm, S., Food Technol. (1967) 21(11), 121-124.
69. Grimes, J. H. and Scott, K. T. B., Powder Met. (1968) 11(22), 213-223.
70. Dress, W. and Grimes, J. H. (United Kingdom Atomic Energy Authority), British Patent 1,175,834, December 23, 1969.
71. Grimes, J. H. and Lane, E. S. (United Kingdom Atomic Energy Authority), British Patent 1,231,385, May 12, 1971.
72. Grimes, J. H. and Lane, E. S. (United Kingdom Atomic Energy Authority), British Patent 1,286,257, August 23, 1972.
73. Grimes, J. H. and Dress, W. (United Kingdom Atomic Energy Authority), British Patent 1,286,871, August 23, 1972.
74. Grimes, J. H., Scott, K. T. B., and McKenna, N. J. (United Kingdom Atomic Energy Authority), British Patent 1,350,389, April 18, 1974.
75. Ball, P. W., Grimes, J. H., and Scott, K. T. B. (United Kingdom Atomic Energy Authority), British Patent 1,420,128, January 7, 1976.
76. Scott, K. T. B., Grimes, J. H., and Ball, P. W. (United Kingdom Atomic Energy Authority), British Patent 1,325,870, August 8, 1973.
77. Scott, K. T. B., Grimes, J. H., and Ball, P. W. (United Kingdom Atomic Energy Authority), British Patent 1,346,295, February 6, 1974.
78. Owen, W. L., Sugar (1950) 45(3), 42-43; Sugar (1955) 50(5), 47-48.
79. Owen, W. L., Sugar (1951) 46(7), 28-30; Sugar (1952) 47(7), 50-51.
80. Owen, W. L., U.S. Patent 2,602,082, July 1, 1952.
81. Ruiz, A. R., Sugar J. (1957) 20(3), 50-52.
82. Richey, Harry and Woods, Jack (Cherokee Laboratories, Tulsa, Oklahoma), personal communications.
83. Mueller, E. P., Z. Angew. Geol. (1963) 9(4), 213-217; Chem Abstr. (1963) 59, 4935a.
84. Monaghan, P. H. and Gidley, J. L., Oil Gas J. (1959) 57(16), 100-103.
85. Heyne, B. and Gabert, A., Bergakademie (1969) 21(5), 285-288; Chem. Abstr. (1969) 71, 62,733r.
86. Sparks, W. J. (Jersey Production Research Co.), U.S. Patent 3,053,765, September 11, 1962.
87. Lindblom, G. P., Ortloff, G. D., and Patton, J. T. (Jersey Production Research Co.), Canadian Patent 654,809, December 25, 1962.
88. Yuen, S., Process Biochem. (November 1974), 7-9.

89. Yuen, S., "Pullulan and Its New Applications," Haya-
    shibara Biochemical Laboratories, Inc., Okayama,
    Japan, February 1974.
90. Zajic, J. E. (Kerr-McGee Oil Industries, Inc.), U.S.
    Patent 3,320,136, May 16, 1967.
91. Goren, M. B. (Kerr-McGee Oil Industries, Inc.), U.S.
    Patent 3,406,114, October 15, 1968.
92. Zajic, J. E. and LeDuy, A., Appl. Microbiol. (1973) 25,
    628-635.
93. Nomura, T. (Sumimoto Chemical Co. Ltd.; Hayashibara
    Biochemical Laboratories, Inc.), U.S. Patent 3,936,347,
    February 3, 1976).

Extracellular Microbial Polysaccharides—A Critical Overview

JEREMY WELLS

Biochem Design S.p.A., Via A. Bargoni, 78, 00153 Rome, Italy

It is the interest of this paper to present a critical overview of commercially significant extracellular microbial polysaccharides within the context of the industrial hydrocolloid or gums market.

Polysaccharide hydrocolloids obtained from plants and seaweed have been used successfully for food, petroleum, textile and numerous industrial applications for several years. Polysaccharides are produced extracellularly by many microorganisms now available. Several of these new hydrocolloids produced microbially, have shown themselves to be commercially significant.

The reasons for the commercial exploitation of these microbial polysaccharides is because of their unique physical and constant chemical properties, regularity of supply, better functional properties and a lower biological use of oxygen.

The commercial usefulness of polysaccharides is based on their ability to alter the rheological properties of water. Present major markets for these polysaccharides exist in the food and the petroleum drilling industries. Large future growth is expected to come from enhanced oil recovery. This paper considers the use of microbial polysaccharides in competition with other water soluble gums in the food industry. It also studies the use of polymers in enhanced oil recovery, when it compares polysaccharides with polyacrylamides.

To provide this overview of the industrial gums markets, it has been necessary to review data recently published or in publication.

Particular thanks are given to Tate and Lyle Ltd. for allowing publication of data recently obtained during a market feasibility study made on their behalf. Particular thanks are given to Dr. C.J.Lawson, without whose cooperation this paper would have been that much more difficult.

THE MARKET FOR WATER SOLUBLE GUMS

Most water soluble gums are theoretically interchangeable. In practice, most gums possess unique characteristics which guarantee their commercial use.

TABLE I
CLASSIFICATION OF NATURAL AND SYNTHETIC
WATER-SOLUBLE GUMS

	Origin	Exemple
	Tree Exudates	Gum Arabic Karaya Gum Gum Tragacanth and others
	Seed Extract	Guar Gum Locust Bean Gum Psyllium and others
	Seaweed Extracts	Agar Alginates Carrageenan and others
Natural	Natural Starches	Corn Starch Potato Starch Tapioca and others
	Natural Products	Dextrans Xanthan Gums Pectin Gelatin
	Starch and Derivates	Dextrins Starch Acetates Dialdehyde star- ches and others
	Cellulose Derivatives	Carboxymethylcel- lulose Methylcellulose Hydroxymethylcel- lulose and other
Synthetic	Petrochemical Derivatives	Polyvinyl Alcohol Polyacrylic Acid Salts Ethylene Oxide Polymers and others

Table I lists both natural and synthetic gums in common use according to class. Gums have diverse chemical composition, origin and functionality and are classified according to their origin.

The main classes of natural gums are the following:

- Natural products   - Tree exudates
- Starch and starch derivates   - Seed extract
- Seaweed extract   - Cellulose derivatives

In the United States, it is reported that while the total expansion of gums is only 1.3% per annum, the synthetic polymer and microbially produced gums are increasing by over 8% per annum.

This increase in consumption of manufactured gums is at the expense of plant gums.

Table II shows the consumption of gums in the United States during 1973 as reported by R.L.Whistler. The overall usage of gums has been fairly widespread throughout the industry.

Originally the tree exudates were the most widely used class of gum. In recent years these exudates have been replaced by manufactured gums including xanthan gum. Improved properties over these plant gums by this range of manufactured gums, have caused the shift. Plant gums vary in quality and are distributed in the raw state, so require further processing including

TABLE II
THE CONSUMPTION OF INDUSTRIAL GUMS IN THE
UNITED STATES (1973)
(tons)

Gum	Food Usage	Industrial Usage	Total Usage
Cornstarch	223,214	1,116,071	1,339,285
Carboxymethylcellulose	6,696	43,303	50,000
Methylcellulose	900	23,660	24,553
Guar	6,696	15,625	22,321
Arabic	10,267	3,125	13,392
Pectin	3,357	0	5,357
Locust bean	4,017	1,785	5,803
Alginate	4,017	4,017	8,034
Ghatti	4,464	446	4,910
Carrageenan	4,017	89	4,106
Xanthan	1,000	2,678	3,678
Karaya	446	3,125	3,571
Tragacanth	580	89	669
Agar	133	178	311
Furcellaran	89	0	89

R.L.Whistler 1974

crushing and cleaning by the end user. While the manufactured gums are in their final state, ready for direct use.

The manufactured gums are often tailor made for specific application or premixed for direct application. These premixes bear a variety of names and are intended for specific purposes.

Seasonal variations in quality, supply and price have often forced processors to change to manufactured gums. However when conditions revert, these processors then prefer to continue to use the reliable manufactured gums. Xanthan for example has taken much of the gum tragacanth market in the U.S.

It is only the food, pharmaceutical and cosmetic industries that still use these plant gums.

The percentage distribution of the manufactured gums in the United States has been reported as follows:

	Percent
Detergents and laundry products	16
Textiles	14
Adhesives	12
Paper	10
Paint	9
Food	8
Pharmaceutical and cosmetic	7
Other	24

The main demands for gums steps from their various functional properties and can be broken down into the following:

Functionality	Percent usage
Stabilizer, suspending agent and dispersant	25
Thickener	23
Film forming agent	17
Water retention agent	12
Coagulant	7
Colloid	6
Lubricant or friction reducer	5
Other purposes	5

Cost effectiveness and cost of gums will be the purchasers most important criteria in deciding which gum to use.

Table III gives a range of prices as reported by the Chemical Market Reporter.

Although gum production is fairly difficult to accurately determine, Table IV lists some gum production estimates obtained from various sources.

The various applications of gums are determined by their cost effectiveness in utilizing physical properties to perform specific applications.

TABLE III

Variety of Gum	Price, $ per lb	
	1971	1975
Agar USP	2.40÷2.80	8.15
Gum Arabic	0.42÷0.60	1.75
Gelatine, edible	0.57÷0.58	1.64÷2.75
Guar gum, edible	0.38÷0.40	0.35÷0.40
Karaya gum	0.80÷0.90	0.90÷0.95
Locust bean gum	0.52÷0.58	0.79÷0.98
Methylcellulose	0.89	0.74
Pectin	2.40	2.22
CMC (carboxymethylcellulose)	0.45	0.60
Gum Tragacanth	2.50÷7.00	10.20÷14.00

Source: International Trade Center, 1972 updated from Chemical
Market Reporter.

TABLE IV
WORLD PRODUCTION OF SELECTED INDUSTRIAL
GUMS

Gum	Year	Production (tons)
Agar 1	1973	7,950
Alginate 1	1973	17,000
Arabic 2	1966	60,000
Carrageenan 1	1973	8,000
Furcellaran 1	1973	1,200
Locust Bean 2	1970	15,000
Methylcellulose 2	1972	25,000
Pectin 2	1971	9,000
Carboxymethylcellulose 2	1969	60,000
Xanthan	1975	5,000

Sources: 1) J.Naylor FAO Production, Trade and Utilization of
Seaweed Products (1976).

2) R.L.Whistler Industrial Gums (1973).

As previously mentioned the main gums are classified into
the following classes:

Natural Products. The four most important hydrocolloids
in this class are gelatin, pectin, dextran and xanthan. In the
food industry over twenty thousand tons of natural polysaccharides were sold in the United States in 1975. This was broken
down into 16,000 tons gelatine, 2,000 tons xanthan and 3,000
tons pectin.

Gelatin is preferred in gelatin desserts, meat products
such as ham and luncheon meat and dairy products. The photographic and pharmaceutical industries are the largest users
of high grade gelatine.

Pectin, because of its gel forming properties with sucrose
is used in jams and confectinery. Xanthan has gained acceptance
in salad dressings, citrus drinks, bakery items and dairy products as well as oil drilling muds and recovery.

Seaweed Extracts. Seaweed extracts are obtained from two
groups of algae, red algae which is the source of carrageenan
and agar and brown algae which is the source of alginates. The
recent FAO Seaweed Resources of the World reports large potential for expansion in this group. Present harvests of red and
brown algae are put at 0.807 and 1.315 million tons respectively with potential outputs listed at 2.66 and 14.6 million tons
of algae.

Each of these gums, has unique properties giving them
excellent market potential. Because of its low gelling and heat
resistance, agar has a wide usage in foods.

Alginates are the most extensively used gum of the group
and are used in dairy products, citrus beverages, bakery fill-
ings, liquid animal feeds, pharmaceutical and many industrial
applications.

Carrageenan has the largest usage of the group in dairy,
beverages and bakery products. It is heavily used in dairy
products because of its reaction with casein.

Starch Derivatives. Natural starches are normally processed to give them properties for special applications. Because
of their strong adhesives properties, they are used in adhesives
and are highly competitive with gum arabic. Other uses include
ceramics, flocculation, well drilling muds and pharmaceuticals.

Seed Extracts. The two major gums in the group are guar
and locust bean gum. Guar is obtained in India and Pakistan,
while locusts bean is harvested in Spain and the Mediterranean
area. They are both used in the dairy industry for cheese making
and ice crean production.

Guar gum is the preferred ice cream stabilizer. Locust bean
gum is a viscosifier and binder of free water.

Cellulose Derivatives. Food, drum, cosmetic and dentifrice
products are the fast growing usage of CMC (sodium carboxy
methylcellulose) and MC (methyl cellulose).

MC costs more than CMC and has a US production of 30 million
pounds against a US production of 74 million pounds of CMC.

Tree Exudates. The tree exudates, gum arabic, gum tragacanth and gum Karaya have all lost market shares to the synthe-
tics and processed gums.

However, gum Karaya because of its laxative properties
maintains a market.

Gum tragacanth has a wide range of applications in food,
textiles, cosmetics and ceramics. However because of supply
difficulties and price differential it has been replaced to a
large extent by xanthan gum.

Gum arabic is used mostly to prevent crystallization of
sugars and as an emulsifier to keep fats uniformly distributed.
CMC, PVA and modified starches have taken a large share of the
gum arabic market.

MICROBIAL POLYSACCHARIDES OF COMMERCIAL SIGNIFICANCE

At this time only three microbial polysaccharides of commercial significance are in commercial production, dextran,
polytran and xanthan. Five others show promise in development
and are awaiting decisions on potential commercial exploitation.

Dextran. Dextrans are polyglucans and have been produced
in the United States, Canada, Holland and Sweden. They can be
synthesised from sucrose microbially from many strains of cell
free culture filtrates of leuconostoc mesenterides, though
dextrans from other strains will differ both in structure and
properties. Molecular weights may range widely. Usual practice
is to obtain a high molecular weight material and degrade it by
hydrolysis, since the dextrans that are used in the food industry must have molecular weights below 100,000, since only they
are included in the GRAS list of the FDA. Dextran solutions are
closely similar to locust bean gum.

Xanthan. Xanthan gum is produced from glucose solution in
growing cultures of xanthomonas campestris.

Commercial production has been carried out in the United
States since 1967 by the Kelco Company, who are currently the
main manufacturer and who produced an estimated 5,000 tons of
xanthan in 1975. Local expansion of the San Diego plant, together with a grass roots plant in Oklahoma for 10,000 tons at
an estimated cost of 35 million dollars plus plant by Rhone
Poulenc of France and General Mills could make available between
35-37½ million pounds xanthan by 1978.

Tate and Lyle Ltd. and Hercules Inc. have announced a
joint-venture to enter in this market.

Current development status of microbial polysaccharides is
listed in Appendix A.

The unique physical properties of xanthan have found many
industrial applications in such diversified industries as textile printing, drilling muds, surfactant flooding, rust removers,
and liquid type of animal feeds.

Most important present use is the recovery of crude oil.
The flow characteristics of xanthan, coupled with its stability

to pH, gives it a technical advantage over other polymers in drilling muds. There is an estimated world usage of 1,800 tons in drilling applications.

Scleroglucan (Polytran). This polysaccharide has been developed by the Pillsbury Company and marketed under the trade name of Polytran. Pillsbury claim important flow characteristics over a wide range of pH and temperature and stability in the presence of salts. Polytran will stabilize bentonite clays during storage, over ranges of temperature and pH. It is used in the ceramic, drilling mud, and inks and coatings industries. Prices in the region of 9-10,000 dollars per ton put it in competition with xanthan gum.

OTHER MICROBIAL POLYSACCHARIDES BEING DEVELOPED

Pullulan. This polysaccharide has been developed by the Hayashabara Company in Japan.

Commercial interest has been shown on account of its ability to form strong resilient films and fibers and the ease it can be molded into shapes. At present pullulan is only in the pilot plant stage.

Construction of a production plant was reported to have started in 1975. Patents claiming both food and industrial applications have been filed.

Microbial Alginate. Alginic acid and alginates are most important gums with many applications in food, textile, pharmaceutical and paper industries. Products obtained from seaweed vary in both, quality and structure. Several microorganisms produce microbial alginates very similar to algal alginate. The composition of these polymers formed is reported to be uneffected by the carbohydrate source used, and of constant quality. Most of the main development in microbial alginates has been claimed by Tate and Lyle.

Curdlan. Curdlan has been developed by the Takeda Company in Japan from a chemical mutant of **Alcaligenes faecalis** var myxogenes 10CS.

Its industrial development depends on the gel strength of high set gels not reverting greatly when cooled. Aqueous suspensions of the polymer remain soft and resilient when cooled, after heating. Applications are immobilized enzymes as well as being used in preparation of films and gel.

Erwina (Zanflo). Erwina was developed by the Kelco Company specially for carpet printing applications, due to its compatibility to cationic dyes. It is produced from a strain of **Erwina tahitica**. It has been claimed that this polysaccharide possesses pseudoplasticity, pH stability and freeze-thaw stability.

The excellent resistance to enzyme attack, and its flow and levelling qualities have already made it find application in the paint industry.

INDUSTRIAL DEVELOPMENT OF MICROBIAL POLYSACCHARIDES.
Dextran was first produced commercially in Sweden in the early forties, then later in England, Canada and the United States.
Xanthan was first produced commercially in the United States in the early sixties for industrial applications. In 1969 FDA cleared the general use of xanthan gum in foods where the standards of identity do not preclude its use. In 1973 FDA allowed uses in process and cream cheeses as a thickening and stabilizing agent. In 1974 MID/PID Inspection Division of USDA included xanthan gums on their authorized list of non meat ingredients.
One can conveniently divide up the main market development of microbial gums into three groups:
- Food applications
- Petroleum and oil industry applications
- Other applications.

Food Applications. Over 60% of microbial polysaccharides sales go to the food industry.
In 1975, Kelco Company are said to have sold over 5 million pounds of xanthan gums into the US food industry. Xanthan gums have gained rapid acceptance into the United States food industry and applications are now being developed in both Europe and Japan. Denmark, England, Ireland, Holland, Spain and Canada have given regulatory approval. Approvals in France, Sweden and Belgium are expected before the end of 1976.
Initially the applications followed in the United States for use in salad dressings, meat analogs, pet food, bakery products, carbonated beverages and frozen foods will be studied. New developments by the USDA, announced during 1975, for producing matrix textures for foods and snack foods could open up very large developments. Joint patents between Kelco Company and DCA give promise of changes in the bakery industry. It is reported that a joint-venture in Japan with this group and Nisshin Flour could open potential markets in donuts, onion ring processing, snack items and certain new bakery products.
In order to see how these microbial polysaccharides fit into the overall market it is necessary to study the total US consumption of gums in 1975, which is given in Table VI. (Xanthan with US sales of 2,500 tons is classified as a natural product.
Initially xanthan gums have obtained their main markets by replacing gum tragacanth. However this market has practically disappeared in the US. Though there exists several thousand tons potential elsewhere.
The world market for alginates is over 17,000 tons with about five thousand tons utilized in the food industry. About 2,000 tons are used in the United States and nearly 1,500 tons in Europe. Largest applications are in dairy and bakery products, where consumption is expected to expand. This could

TABLE VI
US CONSUMPTION OF HYDROCOLLOIDS IN FOODS IN 1975

Product	Thousand tons	Million dollars
Natural products	20.5	113
Starch derivatives	124.2	64
Seaweed extracts	4.1	19
Tree exudates	1.4	7
Seed extracts	5.9	10
Cellulose derivatives	5.9	12
TOTAL	162.0	225

Source: James Hickey - C.H.Kline and Company Inc.

open excellent markets to microbial alginates if they could be
produced at equivalent prices to the algal product.
     In general the shortages of naturally occurring hydro-
colloids in 1974 showed the vulnerability of this market to
less costly manufactured gums. In general the main food
processors will tend to prefer to formulate with these manufactured gums, whose supply, quality and price are not subject
to vagarancies of supply, weather, politics and labour costs.
     The European and Japanese markets are expected to offer
large potential for development. It must be remembered that
over 17,000 tons of xanthan will be available per year after
1978 and over half of this must be absorbed by the Food Industry. Hence the main marketing efforts of xanthan must necessarily move to Europe. Probably a different range of applications will eventually dominate outside the US since use of
salad dressings, meat analogs, and carbonated beverages is not
so developed. For example fruit yoghourt markets are many times
larger than in the US. The largest European hydrocolloid food
usage is in the modified starch field. Over 150,000 tons usage
has been reported to be used in the European Community.
     Probably the best growth rate comes from cellulose deri-
vates. CMC at 60 cents per pound has made great inroads into
the seed gum market. Large volumes of CMC are reported to be
going into instant soups and cake mixes, a market that xanthan
is also trying to penetrate.
     In summary the food ingredients market is very complicated
and only the most technically competent and technically marketed oriented will survive. To sell gums into new food products
demands a sophisticated technical input. It is essential to
understand potential application. Gums are multifunctional.
     Xanthan added to replace an emulsifier, will also increase
viscosity. This can cause problems if the other thickeners are
not reduced.

Petroleum and Oil Industry Applications. Polymers are finding increasing usage in the oil industry and developments are forecast which could open unlimited potential. At this time the market remains in exploration usage. Market developments here has been divided into two segments: oil drilling muds and enhanced oil recovery.

Oil drilling muds. The four major polymers in use at this time are xanthan, polyacrylamides, modified starches and cellulose derivatives particularly CMC. Dextran and pullulan are also trying to get into this industry.

Because of its stability to pH, heat, cations and divalent ions coupled to its pseudoplastic behaviour under conditions of high shear, xanthan gums are the technically preferred polymer for lubrication   of bentonite muds used to drill oil wells.

During 1975 about 1,800 tons of xanthan were used in drilling operations with a potential usage of 3,000 tons predicted by 1980. However recent price increases have caused several of the majors to switch some of the usage to CMC even though it takes almost double the amount of CMC to achieve the same performance effects.

Xanthan is also much used in sumultaneous water flooding and pushing techniques used in the North Sea, where seawater containing small quantities of xanthan (100 ppm) are pushed into injection wells.

However, polyacrylamides are also being considered for this application due to lower price.

The drilling service industry is controlled by a small number of service companies including Baroid, Milchem, Imco, Dresser in the US with Croda and Ceca from Europe. They resell xanthan obtained from Kelco, General Mills or Rhone Poulenc.

Enhanced oil recovery. The greatest future potential for polysaccharides, lies in enhanced oil recovery.

Great interest and research effort is being centered on recovering the large fraction of original oil remaining in place in oil strata after conventional recovery methods have been utilized.

A large incentive for developing recovery enhancement methods exists in the United States as the percentage of imported oil for domestic purposes is increasing very rapidly.

According to the American Petroleum Institute, of the 440 billion barrels of oil discovered in the United States by the end of 1974, 295.8 billion barrels would have to be recovered by enhanced recovery, or advanced enhanced recovery techniques.

In 1973 the Gulf Universities Research Consortium in a study with many of the large oil companies started a series of enhanced recovery. In 1974 very highly optimistic predictions on surfactant flooding gave rise to estimates of recovering 500,000 stock tank barrels (STB) per day. However since

that time, reduced estimates and longer realization times
have been predicted.

At the end of 1975, the consortium were predicting annual
production rates for 1985 of enhanced oil between 300-400
million STB, which would call for an annual polymer demand
of between 200-250 million pounds based on 58% EOR by surfactant flooding.

However in 1976, the Gulf Consortium has how lowered its
1986 prediction to a realistic goal of 200,000 STB per day,
with a starting date for large scale development in 1979.
Table 7 gives the polymer requirements for both cases studied.
This study has assumed that the market is equally shared
between polysaccharides and polyacrylamides.

TABLE VII
POLYMER DEMAND FOR ENHANCED OIL RECOVERY
(Thousands of pounds per day)

Year	Case A	Case B
1979	7.3	18.3
1980	14.6	40.2
1981	32.8	79.0
1982	51.1	131.5
1983	76.7	175.2
1984	11.0	215.2
1985	120.5	233.6
1986	138.5	248.2

Source: Gulf Universities Research Consortium, March 1976
         1. Polymers demand assumed to be 50% polysaccharides,
            50% polyacrylamide.
         2. Case A - assumes development to 200,000 STB/day
                     by 1986.
            Case B - Assumes development to 500,000 STB/day
                     by 1986.

Hence this delphi type exercise predicts markets of between
250 to 140 thousand pounds polymer demand per day by 1986 starting at between 7,300 to 18,300 pounds daily in 1979.

Assuming this potential to be correct, the next problem to
be resolved by those developing polysaccharide polymers, will be
the potential split between polysaccharide and polyacrylamides.
The difficulty in predicting future trends (according to the
Gulf Consortium) is the general dissatisfaction with the current
generation of materials.

Polyacrylamides are in the right price range but are described as unduly shear sensitive and salt sensitive. Table VIII shows
delphi analysis of various figures discussed in numerous studies
since 1974.

TABLE VIII
USE OF POLYACRYLAMIDE AND POLYSACCHARIDE FOR EOR

	Poly-acrylamide	Poly-saccharide	Ideal
Viscosity thickening	10-15 cp at 500 ppm	10-15 cp at 500 ppm	20 cp at 100 ppm
Salinity maximum of fluid	1500-2000 ppm	10000 ppm	15000 ppm
Maximum reservoir temperature	175-200°F	200-225°F	up to 250°F
Divalent ion maximum	200 ppm	5000 ppm	5000 ppm
Permeability oil plugging	Critical>50md	50-100md	>100 md
Cost $/lb	1.30+0.30	2.25+0.30	1.25+2.00

This is based on the Gulf Consortium recommend charge of 10 pounds of surfactants, 3 pounds of alcohols and about 1 pound of polymer per barrel.

Though the use of alcohol is in some doubt, since it may be better to increase the sulphonate ratio at the expense of the alcohol.

It should also be pointed out that surfactant flooding is not the only method of enhanced recovery.

Basic recovery processes have been apportioned as follows:
- Surfactant recovery   58%
- Thermal recovery   29%
- Carbon dioxide processes   8%
-Hydrocarbon miscible processes   5%

A recent development is studying feasibility of developing small scale units to produce surfactant charges including polysaccharides at the oilfield site. It has been noted that EOR has still many technical problems to solve before it will be a commercial reality. However the increasing financial support by Energy Research and Development Administration is most wellcome and indicates political backing which is essential to make this development a reality.

Other Applications. The other applications of microbial polysaccharides have come from taking the market of the natural plant gums with more reliable or tailor made products.

Zanflo has obviously been developed for paint and dying applications. Other polysaccharides have found markets in textiles, cosmetics, pharmaceuticals and liquid feeds. Several patents have recently been issued in Japan for product applications in anticancer preparations.

The industrial uses are more complicated than food uses and are due to rheological properties and wide ranges of stability and compatability with conventional tickening agents and surface active agents.

The synergistic effects of xanthan with locust bean gums is well exploited. However on account of the complexity, most of this know-how remains the confidential property of the processor and supplier of the gum.

CONCLUSION

This paper has tried to establish the place of microbially produced gums in an expanding industrial gum market. It is clear that xanthan gum production will remain at a plateau until the the new product coming into production is absorbed.

The whole industry requires a very high level of technical expertise and marketing skill to develop industrial usage. The future large potentials in enhanced oil recovery are still a long way off and much development work will be required by both the oil producers and polymer suppliers to make this a commercial reality.

Production problems of fermentation drying and reconstituting dilute solution must be solved.

On account of the large development costs necessary for both technical and market development it must be concluded that only companies developing whole range of microbial products will predominate.

APPENDIX "A"
BIOPOLYMER CAPACITY SUMMARY
(Mainly Xanthan)
tons/year)

Company	Affiliates	Location	Capacity	Date
KELCO	Merck subsidiary	San Diego	$3,500^x$	Existing
KELCO	Merck subsidiary	Oklahoma	10,000	End 1976
BIOSYNTHESE-MELLE	Rhone Poulenc General Mills	Melle (France)	2,000	Existing
GENERAL MILLS	Rhone Poulenc	Iowa	2,500	Mid 1977
TATE & LYLE	Hercules Inc.	NA^o	NA^o	NA^o
TATE & LYLE	Hercules Inc.	NA^o	NA^o	NA^o

Total known capacity by end 1979   18,000+ Metric Tons

Estimated "conventional" markets by end 1979   15-16,000 M Tons

x Being expanded to 5,000 t/yr but includes development facilities.

o NA = not announced.

BIBLIOGRAPHY

Arnold C.W. "Chemical Challenges in the Quest for Enhanced Oil
        Recovery" - American Chemical Society (April 4-9
        1976).

Gogarty W.B. "Status of Surfactant or Micellar Methods S.P.E."
        National Meeting, Dallas, Texas (Sept. 28 - Oct. 1
        1975).

Hickey J.R. "Thickeners and Stabilizers for Food ECMRA Meeting
        of Market" - Development Analysts Meeting, London
        (April 1976).

Kang K.S. and Kovacs P. "Int. Congress of Food Science" Madrid
        (1974).

Jeanes A. "Food Technology" (May 1974) 34-39.

Kelco Company "Technical Bulletin DB 21" (1971)

Kimura H. "Abstract 32" Institute of Food Technologists Meeting
        Minneapolis (1972).

Lawson J., Sutherland I.W. "Polysaccharides from Microorganisms
        in Economic Microbiology; ed. by
        A.H. Rose - Academic Press (in press).

"Marketing of Principle Water Soluble Gums in Producing
Countries" International Trade Center, Geneva (1972).

Naylor J. "Production Trade and Utilization of Seaweeds and
        Seaweed Products" FAO, Rome (1976).

"Report on Chemical Demand and Supply Study".

"Relating on Microemulsion Flooding Gulf Universities Research
Consortium" Houston, Texas (March 26, 1976).

Sharp J.M. "The potential of Enhanced Recovery Processes"
        S.P.E. Meeting Dallas, Texas (Sept. 28-Oct. 1 1975)

Umland C.W. "Presentation for Federal Energy Administration"
        Enhanced Oil and Gas Recovery Symposium, Washington
        (Dec. 1975).

Whistler R.L. "Industrial Gums" 2nd Edition Academic Press,
        New York (1973).

INDEX

INDEX

INDEX

317